99 Genetic Engineering Algorithms Handbook With Python

Jamie Flux

https://www.linkedin.com/company/golden-dawn-engineering/

Collaborate with Us!

Have an innovative business idea or a project you'd like to collaborate on?
We're always eager to explore new opportunities for growth and partnership.
Please feel free to reach out to us at:

https://www.linkedin.com/company/golden-dawn-engineering/

We look forward to hearing from you!

Contents

1. Advanced Genome Assembly Algorithms — 7
2. Efficient de Bruijn Graph Construction — 9
3. Sparse Assembly Algorithms for Metagenomics — 12
4. Genome Variation Detection Algorithms — 16
5. Sequence Alignment Optimization Techniques — 18
6. Fast Pairwise Alignment using Seed and Extend Methods — 21
7. Multiple Sequence Alignment with Progressive and Iterative Methods — 24
8. Hidden Markov Models in Gene Prediction — 27
9. Dynamic Programming Algorithms in RNA Secondary Structure Prediction — 30
10. Machine Learning Models for Variant Calling — 33
11. Deep Learning for Functional Genomics — 36
12. Graph-based Genomic Data Structures — 39
13. Phylogenetic Tree Construction Algorithms — 42
14. Maximum Likelihood Estimation in Phylogenetics — 44
15. Bayesian Inference in Evolutionary Biology — 47

16	Coalescent Theory Algorithms	50
17	Genome-Wide Association Study (GWAS) Algorithms	52
18	Linkage Disequilibrium Mapping Techniques	55
19	Population Genetics Simulation Algorithms	58
20	Metabolic Pathway Modeling and Simulation	61
21	Flux Balance Analysis in Metabolic Engineering	64
22	Constraint-Based Reconstruction and Analysis (COBRA) Methods	67
23	Synthetic Gene Circuit Design Algorithms	70
24	Boolean Network Modeling of Gene Regulatory Networks	72
25	Stochastic Simulations of Cellular Processes	75
26	Agent-Based Modeling in Systems Biology	78
27	Protein Structure Prediction with Homology Modeling	81
28	Molecular Dynamics Simulation Algorithms	84
29	Protein Docking Algorithms	87
30	Cryo-EM Data Processing Algorithms	90
31	Optimizing CRISPR Guide RNA Design	93
32	Off-Target Prediction in Genome Editing	96
33	Homology-Directed Repair Simulation	99
34	Gene Expression Quantification Algorithms	101
35	Single-Cell RNA Sequencing Data Analysis	104
36	Clustering Techniques for Cellular Heterogeneity	107

37 Dimensionality Reduction in Genomic Data	110
38 Gene Regulatory Network Inference	113
39 Transcription Factor Binding Site Prediction	116
40 Chromatin Accessibility Analysis	118
41 Epigenetic Modification Modeling	121
42 Non-Coding RNA Function Prediction	124
43 miRNA Target Prediction Algorithms	127
44 Long Non-Coding RNA Interaction Networks	130
45 Sequence Motif Discovery Algorithms	133
46 Genome Annotation Pipelines	136
47 Functional Annotation of Genes and Proteins	139
48 Pathway Enrichment Analysis	142
49 Comparative Genomics Algorithms	145
50 Ortholog and Paralog Identification	148
51 Structural Variation Detection Algorithms	151
52 Gene Fusion Detection in Cancer Genomics	154
53 Protein-Protein Interaction Networks	157
54 Network Topology Analysis in Systems Biology	160
55 Metagenomic Data Analysis Algorithms	163
56 Taxonomic Classification in Metagenomics	166
57 Functional Profiling of Microbial Communities	169
58 Genome Reconstruction from Metagenomic Data	172
59 Gene Ontology Enrichment Analysis	175

60 Antibody Design Algorithms	178
61 Computational Vaccine Design	181
62 In Silico Drug Discovery Platforms	184
63 Gene Therapy Vector Optimization	187
64 Algorithmic Approaches to Gene Synthesis	190
65 Codon Usage Bias Analysis	193
66 Error Correction Algorithms in DNA Synthesis	196
67 High-Throughput Sequencing Error Correction	199
68 Data Compression Techniques for Genomics	202
69 Secure Sharing of Genomic Data	205
70 Integration of Multi-Omics Data	207
71 Statistical Methods in Bioinformatics	210
72 Bayesian Networks in Genetic Analysis	213
73 Markov Chain Monte Carlo Methods in Genetics	215
74 High-Performance Computing for Genomics	218
75 Cloud Computing Architectures for Bioinformatics	221
76 GPU Acceleration of Bioinformatics Algorithms	224
77 Reinforcement Learning in Genetic Engineering	227
78 Evolutionary Algorithms for Protein Engineering	230
79 Genetic Programming in Synthetic Biology	233
80 Optimization Algorithms in Metabolic Engineering	237
81 Swarm Intelligence in Biological Systems	240
82 Bioinformatics Workflow Automation	244

83	Semantic Web in Genomics	247
84	Ontology Development for Genetic Data	249
85	Natural Language Processing in Genomics	252
86	Automated Hypothesis Generation in Biology	255
87	Text Mining for Genetic Associations	258
88	Visualization Techniques for Genomic Data	261
89	Interactive Genome Browsers Development	264
90	Algorithmic Music Generation from Genetic Data	267
91	Fractal Patterns in Genomic Sequences	269
92	Chaos Theory Applications in Genetics	271
93	Multi-Objective Optimization in Genetic Engineering	273
94	Adaptive Algorithms in Evolutionary Biology	276
95	Information Theory in Genetics	280
96	Game Theory Models in Genetic Interactions	282
97	Artificial Life and Genetic Algorithms	285
98	Synthetic Evolution and In Silico Selection	288
99	Cellular Automata Models in Genomic Pattern Formation	292

Chapter 1

Advanced Genome Assembly Algorithms

Below is a Python code snippet that encompasses the core computational elements necessary for assembling large and complex genomes using high-throughput sequencing data. It includes the implementation of de Bruijn graph construction from sequence reads.

```
from collections import defaultdict

def build_kmers(sequence, k):
    '''
    Generate k-mers from a sequence.
    :param sequence: DNA sequence string.
    :param k: Length of the k-mer.
    :return: List of k-mers.
    '''
    return [sequence[i:i + k] for i in range(len(sequence) - k + 1)]

def build_de_bruijn_graph(sequences, k):
    '''
    Build a de Bruijn graph from a list of sequences.
    :param sequences: A list of DNA sequences.
    :param k: Length of the k-mer (overlap of k-1).
    :return: A dictionary representing the adjacency list of the de
    ↪ Bruijn graph.
    '''
    edges = defaultdict(list)
    for sequence in sequences:
        kmers = build_kmers(sequence, k)
        for i in range(len(kmers) - 1):
```

```
            prefix = kmers[i]
            suffix = kmers[i + 1]
            edges[prefix].append(suffix)
    return edges

def assemble_genome_from_de_bruijn(graph):
    '''
    Perform Eulerian walk to assemble the genome from de Bruijn
    ↪ graph.
    This is a simplified algorithm for genome assembly.
    :param graph: The adjacency list of the de Bruijn graph.
    :return: Assembled genome sequence.
    '''
    start = list(graph.keys())[0]
    path = []
    stack = [start]

    while stack:
        node = stack[-1]
        if graph[node]:
            next_node = graph[node].pop()
            stack.append(next_node)
        else:
            path.append(stack.pop())

    return ''.join(p[0] for p in path[:-1]) + path[-1]

# Example usage
sequences = ["ACGTAGCTAG", "CTAGCGTACG", "GCTAGCTA"]
k = 4
graph = build_de_bruijn_graph(sequences, k)
assembled_genome = assemble_genome_from_de_bruijn(graph)

print("Assembled Genome:", assembled_genome)
```

This code defines several key functions necessary for genome assembly algorithms:

- `build_kmers` function generates k-mers from a given DNA sequence for a given k-mer length.

- `build_de_bruijn_graph` creates a de Bruijn graph from the sequences to represent overlaps between k-mers.

- `assemble_genome_from_de_bruijn` implements a simplified Eulerian walk algorithm to reconstruct the genome from the de Bruijn graph.

The code includes an example usage to demonstrate the process of building the graph from the sequences and assembling the genome.

Chapter 2

Efficient de Bruijn Graph Construction

Below is a Python code snippet that encapsulates the primary computational components for constructing and optimizing de Bruijn graphs, pivotal for enhancing genome assembly efficiency and accuracy.

```
class DeBruijnGraph:
    '''
    Represents a de Bruijn graph for sequence assembly.
    '''
    def __init__(self, k):
        '''
        Initialize the graph with k-mer length.
        :param k: Length of k-mer.
        '''
        self.k = k
        self.edges = {}
        self.nodes = set()

    def add_kmer(self, kmer):
        '''
        Add a k-mer to the graph.
        :param kmer: A k-length string.
        '''
        prefix = kmer[:-1]
        suffix = kmer[1:]
        self.nodes.add(prefix)
        self.nodes.add(suffix)
        if prefix in self.edges:
            self.edges[prefix].append(suffix)
```

```python
        else:
            self.edges[prefix] = [suffix]

    def construct(self, sequence):
        '''
        Construct the de Bruijn graph from the input sequence.
        :param sequence: A DNA sequence string.
        '''
        for i in range(len(sequence) - self.k + 1):
            kmer = sequence[i:i + self.k]
            self.add_kmer(kmer)

    def eulerian_path(self):
        '''
        Find an Eulerian path in the constructed graph.
        :return: List of nodes representing the Eulerian path.
        '''
        in_degree = {node: 0 for node in self.nodes}
        out_degree = {node: len(self.edges.get(node, [])) for node
        ↪   in self.nodes}

        for node in self.edges:
            for adjacent in self.edges[node]:
                in_degree[adjacent] += 1

        start_node = None
        for node in self.nodes:
            if out_degree[node] - in_degree[node] == 1:
                start_node = node
                break

        path = []
        stack = [start_node or self.nodes.pop()]
        while stack:
            while self.edges.get(stack[-1], []):
                stack.append(self.edges[stack[-1]].pop())
            path.append(stack.pop())

        return path[::-1]

    def assemble_sequence(self):
        '''
        Reconstruct the sequence from the Eulerian path.
        :return: A string representing the assembled sequence.
        '''
        path = self.eulerian_path()
        sequence = path[0]
        for node in path[1:]:
            sequence += node[-1]
        return sequence

# Example usage
sequence = "AAGATTCTCTAAGA"
```

```
k = 4

graph = DeBruijnGraph(k)
graph.construct(sequence)
assembled = graph.assemble_sequence()
print("Assembled sequence:", assembled)
```

This code encapsulates vital functions necessary for the creation and optimization of de Bruijn graphs:

- The `DeBruijnGraph` class models a graph where nodes represent (k-1)-mers, and edges correspond to k-mers in the genomic sequence.

- `add_kmer` integrates individual k-mers into the graph by connecting their prefixes and suffixes.

- `construct` generates the entire de Bruijn graph from a given DNA sequence by iterating through and adding each k-mer.

- `eulerian_path` computes a Eulerian path, formulating the basis for reconstructing sequences from the graph.

- `assemble_sequence` utilizes the Eulerian path to output the contiguous sequence derived from the assembly process.

An example is provided to illustrate constructing a de Bruijn graph for a sequence and assembling it back into a sequence.

Chapter 3

Sparse Assembly Algorithms for Metagenomics

Below is a Python code snippet that encompasses the core computational elements for assembling genomes from sparse and complex metagenomic datasets, focusing on an assembly graph approach, error handling, and scaffold construction.

```python
import networkx as nx
from Bio import SeqIO
from collections import defaultdict

def parse_reads(file_path):
    '''
    Parse the sequence reads from a file.
    :param file_path: Path to the fasta file containing reads.
    :return: List of sequence reads.
    '''
    reads = []
    for record in SeqIO.parse(file_path, "fasta"):
        reads.append(str(record.seq))
    return reads

def build_de_bruijn_graph(reads, k):
    '''
    Construct a de Bruijn graph from sequence reads.
    :param reads: List of sequence reads.
    :param k: Length of k-mer.
    :return: de Bruijn graph.
    '''
```

```python
        graph = nx.DiGraph()
        kmers = defaultdict(int)

        for read in reads:
            for i in range(len(read) - k + 1):
                kmer = read[i:i + k]
                next_kmer = read[i + 1:i + k + 1]
                if next_kmer:
                    kmers[(kmer, next_kmer)] += 1

        for (kmer, next_kmer), count in kmers.items():
            graph.add_edge(kmer, next_kmer, weight=count)

        return graph

    def remove_tips(graph, min_tip_length=2):
        '''
        Remove short tips from the de Bruijn graph.
        :param graph: de Bruijn graph.
        :param min_tip_length: Minimum length of tips to be retained.
        :return: Graph with tips removed.
        '''
        nodes_to_remove = []

        for node in list(graph.nodes):
            if graph.in_degree(node) == 0 or graph.out_degree(node) ==
            ↪ 0:
                path_length = 0
                current_node = node

                while path_length < min_tip_length:
                    neighbors = list(graph.successors(current_node))
                    if len(neighbors) != 1:
                        break
                    current_node = neighbors[0]
                    path_length += 1

                if path_length < min_tip_length:
                    nodes_to_remove.append(node)

        graph.remove_nodes_from(nodes_to_remove)
        return graph

    def assemble_contigs(graph):
        '''
        Assemble contigs from the de Bruijn graph.
        :param graph: de Bruijn graph.
        :return: List of assembled contigs.
        '''
        contigs = []
        for component in nx.weakly_connected_components(graph):
            subgraph = graph.subgraph(component).copy()
            path = []
```

```
        while subgraph:
            start_node = max(subgraph.degree, key=lambda x: x[1])[0]
            traverse_path = dfs_with_merge(subgraph, start_node)
            if traverse_path:
                contigs.append(traverse_path)
                subgraph.remove_nodes_from(traverse_path)

    return contigs

def dfs_with_merge(graph, start):
    '''
    Perform a DFS traversal merging nodes into contigs.
    :param graph: Subgraph of de Bruijn graph.
    :param start: Starting node for BFS.
    :return: Merged contig.
    '''
    path = [start]
    current_node = start

    while graph.out_degree(current_node) == 1:
        successor = list(graph.successors(current_node))[0]
        path.append(successor[-1])  # append last character of the
        ↪ successor
        current_node = successor

    return ''.join(path)

# Example usage
reads = parse_reads("metagenomic_reads.fasta")
de_bruijn_graph = build_de_bruijn_graph(reads, k=31)
cleaned_graph = remove_tips(de_bruijn_graph, min_tip_length=2)
contigs = assemble_contigs(cleaned_graph)

print("Assembled contigs:", contigs)
```

This code defines several key functions necessary for metagenomic genome assembly:

- `parse_reads` reads sequence files and loads them for processing.

- `build_de_bruijn_graph` constructs a de Bruijn graph using k-mers extracted from the input reads.

- `remove_tips` removes short, likely erroneous branches (tips) from the graph to improve assembly quality.

- `assemble_contigs` extracts linear sequences (contigs) from the refined graph for further analysis.

- `dfs_with_merge` performs depth-first traversal, merging sequences into full contigs.

This approach processes sparse metagenomic reads, helping to resolve complex species mixtures by constructing and simplifying the assembly graph iteratively.

Chapter 4

Genome Variation Detection Algorithms

Below is a Python code snippet that demonstrates the implementation of advanced methods for detecting single nucleotide polymorphisms (SNPs), insertions, deletions, and structural variations in genomic data.

```
import pysam
import numpy as np

def detect_variants(bam_file, reference_fasta):
    '''
    Detects SNPs, insertions, deletions, and structural variations
    in genomic data.
    :param bam_file: Path to the BAM file with aligned reads.
    :param reference_fasta: Path to the reference genome FASTA file.
    :return: List of detected variants.
    '''
    samfile = pysam.AlignmentFile(bam_file, "rb")
    reference = pysam.FastaFile(reference_fasta)
    variants = []

    for pileupcolumn in samfile.pileup():
        ref_base =
            reference.fetch(reference=char(pileupcolumn.reference_id),
                            start=pileupcolumn.pos,
                            end=pileupcolumn.pos + 1)
        base_counts = {"A": 0, "C": 0, "G": 0, "T": 0, "N": 0}

        for pileupread in pileupcolumn.pileups:
            if not pileupread.is_del and not pileupread.is_refskip:
                base = pileupread.alignment.
```

```python
                        query_sequence[pileupread.query_position]
                        base_counts[base] += 1

            total_count = sum(base_counts.values()) - base_counts["N"]
            if total_count > 0:
                for base, count in base_counts.items():
                    if base != ref_base and count / total_count > 0.2:
                        # 20% threshold for variation
                        variants.append({
                            'position': pileupcolumn.pos,
                            'reference': ref_base,
                            'variant': base,
                            'frequency': count / total_count
                        })

    samfile.close()
    reference.close()
    return variants

def main():
    bam_file = "example.bam"
    reference_fasta = "reference.fasta"
    variants = detect_variants(bam_file, reference_fasta)

    for variant in variants:
        print(f"Position: {variant['position']}, "
              f"Ref: {variant['reference']}, "
              f"Variant: {variant['variant']}, "
              f"Frequency: {variant['frequency']:.2f}")

if __name__ == "__main__":
    main()
```

This code includes functions necessary for detecting genetic variations:

- `detect_variants` is the main function responsible for processing BAM files and identifying variations against a reference genome.

- It uses the `pysam` library to efficiently read and manage genomic data.

- The method identifies variants by comparing read-aligned sequences to the reference sequence and calculating a frequency threshold for calling a variant.

The provided example showcases processing a BAM file to detect and print variants with a frequency above a specified threshold.

Chapter 5

Sequence Alignment Optimization Techniques

Below is a Python code snippet that demonstrates optimizing sequence alignment algorithms for speed and accuracy in large-scale genomic data analysis. This implementation features key components of the Needleman-Wunsch algorithm with optimizations for handling large datasets efficiently.

```python
import numpy as np

def needleman_wunsch(seq1, seq2, match=1, mismatch=-1, gap=-1):
    '''
    Perform global sequence alignment using Needleman-Wunsch
    ↪ algorithm.
    :param seq1: First sequence to align.
    :param seq2: Second sequence to align.
    :param match: Score for matching characters.
    :param mismatch: Score for mismatching characters.
    :param gap: Score for a gap.
    :return: Aligned sequences and alignment score.
    '''
    n = len(seq1)
    m = len(seq2)

    # Initialize scoring matrix
    score_matrix = np.zeros((n + 1, m + 1))

    # Initialize first row and column
```

```python
    for i in range(1, n + 1):
        score_matrix[i][0] = i * gap
    for j in range(1, m + 1):
        score_matrix[0][j] = j * gap

    # Fill scoring matrix
    for i in range(1, n + 1):
        for j in range(1, m + 1):
            match_score = score_matrix[i-1][j-1] + (match if
                seq1[i-1] == seq2[j-1] else mismatch)
            delete_score = score_matrix[i-1][j] + gap
            insert_score = score_matrix[i][j-1] + gap
            score_matrix[i][j] = max(match_score, delete_score,
                insert_score)

    # Traceback to get alignments
    align1, align2 = '', ''
    i, j = n, m

    while i > 0 and j > 0:
        current_score = score_matrix[i][j]
        if current_score == score_matrix[i-1][j-1] + (match if
            seq1[i-1] == seq2[j-1] else mismatch):
            align1 += seq1[i-1]
            align2 += seq2[j-1]
            i -= 1
            j -= 1
        elif current_score == score_matrix[i-1][j] + gap:
            align1 += seq1[i-1]
            align2 += '-'
            i -= 1
        else:
            align1 += '-'
            align2 += seq2[j-1]
            j -= 1

    # Fill remaining elements if any
    while i > 0:
        align1 += seq1[i-1]
        align2 += '-'
        i -= 1
    while j > 0:
        align1 += '-'
        align2 += seq2[j-1]
        j -= 1

    return align1[::-1], align2[::-1], score_matrix[n][m]

# Example sequences
sequence1 = "GATTACA"
sequence2 = "GCATGCU"

# Perform alignment
```

```
aligned_seq1, aligned_seq2, score = needleman_wunsch(sequence1,
↪   sequence2)

print("Aligned Sequence 1:", aligned_seq1)
print("Aligned Sequence 2:", aligned_seq2)
print("Alignment Score:", score)
```

This code elucidates the core components of optimizing sequence alignments:

- `needleman_wunsch` performs global sequence alignment using the Needleman-Wunsch algorithm with parameters for match, mismatch, and gap penalties.

- The scoring matrix is initialized, scored, and then traced back to derive the optimal alignment.

- The function returns the aligned sequences and their alignment score, which can be used to inform further analysis or decision-making.

By applying this algorithm to example sequences, users can gain insights into both the sequence alignment process and its applications in genomic data analysis.

Chapter 6

Fast Pairwise Alignment using Seed and Extend Methods

Below is a Python code snippet implementing the core computational elements of the seed and extend algorithm for accelerating pairwise sequence alignment. This approach involves anchoring short "seed" matches followed by extending these matches to generate optimal alignments.

```python
import numpy as np

def find_seeds(query, subject, seed_length):
    '''
    Find exact matches of a given seed length between query and
    ↪ subject.
    :param query: The query sequence.
    :param subject: The subject sequence.
    :param seed_length: Length of the seeds.
    :return: List of seed matches as tuples (query_index,
    ↪ subject_index).
    '''
    seed_matches = []
    query_len = len(query)
    subject_len = len(subject)

    for i in range(query_len - seed_length + 1):
        seed = query[i:i + seed_length]
        for j in range(subject_len - seed_length + 1):
            if subject[j:j + seed_length] == seed:
```

```python
            seed_matches.append((i, j))

    return seed_matches

def extend_seed(query, subject, query_index, subject_index,
    seed_length):
    '''
    Extend seed matches to longer alignments with scoring.
    :param query: The query sequence.
    :param subject: The subject sequence.
    :param query_index: Start index in query for seed.
    :param subject_index: Start index in subject for seed.
    :param seed_length: Length of the seed.
    :return: Alignment score for the extended region.
    '''
    score = 0
    i, j = query_index, subject_index

    # Extend forward
    while i < len(query) and j < len(subject) and query[i] ==
        subject[j]:
        score += 1
        i += 1
        j += 1

    # Extend backward
    i, j = query_index - 1, subject_index - 1
    while i >= 0 and j >= 0 and query[i] == subject[j]:
        score += 1
        i -= 1
        j -= 1

    return score

def calculate_alignment(query, subject, seed_length):
    '''
    Calculate the best alignment score using seed and extend.
    :param query: The query sequence.
    :param subject: The subject sequence.
    :param seed_length: Length of the seeds.
    :return: Best alignment score.
    '''
    seeds = find_seeds(query, subject, seed_length)
    best_score = 0

    for query_index, subject_index in seeds:
        score = extend_seed(query, subject, query_index,
            subject_index, seed_length)
        if score > best_score:
            best_score = score

    return best_score
```

```
# Example sequences and seed length
query_sequence = "ACGTACGTGAC"
subject_sequence = "TACGTACGT"
seed_len = 3

# Compute the alignment score
alignment_score = calculate_alignment(query_sequence,
↪    subject_sequence, seed_len)
print("Best Alignment Score:", alignment_score)
```

This code defines several key functions necessary for the implementation of the seed and extend algorithm for sequence alignment:

- `find_seeds` identifies exact matches of a specified length (seeds) between two sequences.

- `extend_seed` takes these seed matches and extends them by scoring additional matching bases before and after the seeds.

- `calculate_alignment` computes the best alignment score by utilizing the identified seeds and performing extensions.

The final block of code shows how to compute the alignment score using sample sequences and a defined seed length.

Chapter 7

Multiple Sequence Alignment with Progressive and Iterative Methods

Below is a Python code snippet that demonstrates multiple sequence alignment using the `progressive alignment` and `iterative refinement` techniques. This example utilizes the `Biopython` library for basic operations and outlines how these techniques can be implemented to align sequences effectively.

```
from Bio.Seq import Seq
from Bio.Align import MultipleSeqAlignment
from Bio.Align.substitution_matrices import Array, blosum62

def calculate_substitution_matrix(seq1, seq2):
    '''
    Calculate a substitution score between two sequences using
    ↪ BLOSUM62.
    :param seq1: First sequence.
    :param seq2: Second sequence.
    :return: Score of substitution.
    '''
    score = 0
    for a, b in zip(seq1, seq2):
        score += blosum62[(a, b)]
    return score
```

```python
def initial_alignment(sequences):
    '''
    Perform initial progressive alignment.
    :param sequences: List of sequences to align.
    :return: Initial multiple sequence alignment.
    '''
    alignments = []
    # Start with first two sequences
    aligned_seq1 = sequences[0]
    aligned_seq2 = sequences[1]
    progressive_align(aligned_seq1, aligned_seq2, alignments)
    # Continue adding sequences iteratively
    for seq in sequences[2:]:
        align_with_existing(alignments, seq)
    return MultipleSeqAlignment(alignments)

def progressive_align(seq1, seq2, alignments):
    '''
    Align two sequences using progressive alignment technique.
    :param seq1: First sequence
    :param seq2: Second sequence
    :param alignments: Alignment object container
    '''
    score = calculate_substitution_matrix(seq1, seq2)
    if score > 0:
        alignments.append(seq1 + '-' * (len(seq2) - len(seq1)))
        alignments.append(seq2 + '-' * (len(seq1) - len(seq2)))

def align_with_existing(alignments, new_seq):
    '''
    Align a new sequence with existing alignments iteratively.
    :param alignments: Existing alignments
    :param new_seq: New sequence to align
    '''
    best_alignment = None
    max_score = -float('inf')
    for seq in alignments:
        score = calculate_substitution_matrix(seq, new_seq)
        if score > max_score:
            max_score = score
            best_alignment = seq

    adjusted_alignment = adjust_to_best_fit(best_alignment, new_seq)
    alignments.append(adjusted_alignment)

def adjust_to_best_fit(aligned_seq, new_seq):
    '''
    Adjust new sequence to fit best existing alignment.
    :param aligned_seq: Best existing alignment
    :param new_seq: New sequence
    :return: Adjusted sequence
    '''
    return new_seq + '-' * (len(aligned_seq) - len(new_seq))
```

```python
# Example of sequence alignment
seq1 = Seq("MKVIL")
seq2 = Seq("MKL")
seq3 = Seq("MVIL")
initial_seqs = [seq1, seq2, seq3]

alignment = initial_alignment(initial_seqs)
for record in alignment:
    print(record)
```

This Python code snippet implements a basic framework for progressive and iterative sequence alignment using the following key components:

- `calculate_substitution_matrix` computes a substitution score based on the BLOSUM62 matrix to gauge alignment quality.

- `initial_alignment` starts the process with initial progressive alignment, progressively adding new sequences.

- `progressive_align` handles pairwise alignment and integration of initial sequences into the alignment framework.

- `align_with_existing` adds sequences iteratively by aligning them to previous alignments and maximizing substitution scores.

- `adjust_to_best_fit` modifies new sequences to match the length of the best-aligned sequence, ensuring structural consistency.

This approach illustrates the conceptual basis of multiple sequence alignment using simplified illustrations and is adaptable for more complex real-world applications.

Chapter 8

Hidden Markov Models in Gene Prediction

Below is a Python code snippet that encompasses the core computational elements of applying Hidden Markov Models (HMMs) for accurate gene prediction in genomic sequences. This includes the initialization of model parameters, implementation of the Viterbi algorithm for state sequence prediction, and decoding the optimal path corresponding to gene annotations.

```python
import numpy as np

def initialize_hmm_parameters(states, observations):
    '''
    Initialize the parameters for the HMM.
    :param states: List of states.
    :param observations: List of observations.
    :return: Initial probabilities, transition probabilities, and
    ↪    emission probabilities.
    '''
    # Initial probabilities
    init_probs = {state: 1.0 / len(states) for state in states}

    # Transition probabilities
    trans_probs = {state: {next_state: 1.0 / len(states) for
    ↪    next_state in states} for state in states}

    # Emission probabilities
    emit_probs = {state: {obs: 1.0 / len(observations) for obs in
    ↪    observations} for state in states}

    return init_probs, trans_probs, emit_probs
```

```python
def viterbi_algorithm(observations, states, init_probs, trans_probs,
    ↪ emit_probs):
    '''
    Implementation of the Viterbi algorithm to find the most
    ↪ probable state sequence.
    :param observations: Sequence of observations.
    :param states: Set of states.
    :param init_probs: Initial probabilities.
    :param trans_probs: Transition probabilities.
    :param emit_probs: Emission probabilities.
    :return: Most probable state sequence.
    '''
    # Length of observation sequence
    T = len(observations)

    # Initialization
    V = np.zeros((len(states), T))   # Viterbi matrix
    path = np.zeros((len(states), T), dtype=int)   # Path to states

    # Map states to indexes
    state_index = {state: i for i, state in enumerate(states)}

    # Initial probabilities
    for state in states:
        V[state_index[state], 0] = init_probs[state] *
        ↪ emit_probs[state][observations[0]]

    # Dynamic programming forward pass
    for t in range(1, T):
        for state in states:
            max_prob = max(
                V[state_index[prev_state], t-1] *
                ↪ trans_probs[prev_state][state] for prev_state in
                ↪ states
            )
            V[state_index[state], t] = max_prob *
            ↪ emit_probs[state][observations[t]]

            # Store the path of the best state transition
            path[state_index[state], t] = np.argmax([
                V[state_index[prev_state], t-1] *
                ↪ trans_probs[prev_state][state] for prev_state in
                ↪ states
            ])

    # Backtracking to get the most probable state path
    best_path = np.zeros(T, dtype=int)
    best_path[T-1] = np.argmax(V[:, T-1])

    for t in reversed(range(T-1)):
        best_path[t] = path[best_path[t+1], t+1]
```

```
# Map indexes back to states
inv_state_index = {v: k for k, v in state_index.items()}
best_state_sequence = [inv_state_index[state_index] for
↪    state_index in best_path]

return best_state_sequence

# Example usage
states = ['Exon', 'Intron']
observations = ['A', 'C', 'G', 'T']
init_probs, trans_probs, emit_probs =
↪    initialize_hmm_parameters(states, observations)

# Assume this is our observed sequence of nucleotides
observation_sequence = ['G', 'C', 'A', 'T', 'A', 'G', 'G', 'C', 'T']

# Decode the most probable state sequence
predicted_states = viterbi_algorithm(observation_sequence, states,
↪    init_probs, trans_probs, emit_probs)

print("Predicted States:", predicted_states)
```

This code defines several key functions necessary for implementing Hidden Markov Models for gene prediction:

- `initialize_hmm_parameters` sets up the initial probabilities, transition probabilities, and emission probabilities for the HMM given a set of states and observations.

- `viterbi_algorithm` utilizes dynamic programming to decode the most probable sequence of states (such as 'Exon' and 'Intron') based on observed sequences of nucleotides.

The final block of code demonstrates how to process an observed sequence of nucleotides to predict the most likely sequence of genomic states using the Viterbi algorithm.

Chapter 9

Dynamic Programming Algorithms in RNA Secondary Structure Prediction

Below is a Python code snippet that encompasses the dynamic programming approach to RNA secondary structure prediction. It utilizes algorithms that calculate the optimal folding of an RNA sequence to achieve a minimal free energy configuration.

```
import numpy as np

def initialize_matrix(n):
    '''
    Initialize a score matrix for RNA secondary structure
    ↪ prediction.
    :param n: Length of the RNA sequence.
    :return: Initialized matrix with zeroes.
    '''
    return np.zeros((n, n))

def is_complementary(base1, base2):
    '''
    Determine if two RNA bases can form a Watson-Crick pair.
    :param base1: First base.
    :param base2: Second base.
    :return: Boolean indicating if bases pair.
    '''
```

```python
    pairs = {'A': 'U', 'U': 'A', 'C': 'G', 'G': 'C'}
    return pairs.get(base1) == base2

def fill_matrix(matrix, sequence):
    '''
    Fill the dynamic programming matrix for RNA folding.
    :param matrix: DP matrix to be filled.
    :param sequence: RNA sequence.
    :return: Filled matrix with computed scores.
    '''
    n = len(sequence)
    for length in range(1, n):
        for i in range(n - length):
            j = i + length
            pair_score = matrix[i + 1][j - 1] + (1 if
             ↪  is_complementary(sequence[i], sequence[j]) else 0)
            unpair_score = max(matrix[i + 1][j], matrix[i][j - 1])
            matrix[i][j] = max(pair_score, unpair_score)
            for k in range(i + 1, j):
                matrix[i][j] = max(matrix[i][j], matrix[i][k] +
                 ↪  matrix[k + 1][j])
    return matrix

def backtrack(matrix, sequence, i, j):
    '''
    Backtrack the matrix to retrieve the optimal RNA folding.
    :param matrix: DP matrix to backtrack.
    :param sequence: RNA sequence.
    :param i: Start index.
    :param j: End index.
    :return: List of base pairs in the folded structure.
    '''
    if i >= j:
        return []
    if is_complementary(sequence[i], sequence[j]) and matrix[i][j]
     ↪  == matrix[i + 1][j - 1] + 1:
        return [(i, j)] + backtrack(matrix, sequence, i + 1, j - 1)
    if matrix[i][j] == matrix[i + 1][j]:
        return backtrack(matrix, sequence, i + 1, j)
    if matrix[i][j] == matrix[i][j - 1]:
        return backtrack(matrix, sequence, i, j - 1)
    for k in range(i + 1, j):
        if matrix[i][j] == matrix[i][k] + matrix[k + 1][j]:
            return backtrack(matrix, sequence, i, k) +
             ↪  backtrack(matrix, sequence, k + 1, j)
    return []

def predict_structure(sequence):
    '''
    Predict the optimal secondary structure for an RNA sequence.
    :param sequence: RNA sequence.
    :return: List of base pairs representing the predicted
     ↪  structure.
```

```
'''
n = len(sequence)
matrix = initialize_matrix(n)
filled_matrix = fill_matrix(matrix, sequence)
return backtrack(filled_matrix, sequence, 0, n - 1)

# Example usage
rna_sequence = "GCGGAUUUAGCUCAGUUGG"
predicted_pairs = predict_structure(rna_sequence)
print("Predicted Base Pairs:", predicted_pairs)
```

This code defines several key functions necessary for RNA secondary structure prediction:

- `initialize_matrix` function sets up a matrix used for dynamic programming, initialized with zeroes.

- `is_complementary` checks if two given RNA bases can form a valid Watson-Crick pairing.

- `fill_matrix` computes the maximum number of base pairs possible using dynamic programming.

- `backtrack` retrieves the optimal pairing path from the computed matrix.

- `predict_structure` utilizes the previous functions to predict and return the optimal secondary structure for an RNA sequence.

The final block of code demonstrates predicting base pairs in an example RNA sequence using the dynamic programming approach.

Chapter 10

Machine Learning Models for Variant Calling

Below is a Python code snippet that showcases the integration of machine learning algorithms to enhance variant detection accuracy in genomic data. This implementation utilizes a simple neural network to predict genomic variants and demonstrates the training process using synthetic data.

```python
import numpy as np
import tensorflow as tf
from tensorflow.keras.models import Sequential
from tensorflow.keras.layers import Dense, Dropout
from sklearn.model_selection import train_test_split
from sklearn.preprocessing import StandardScaler

# Generate synthetic genomic data
def generate_synthetic_data(num_samples=1000, num_features=50):
    X = np.random.rand(num_samples, num_features)
    y = np.random.randint(2, size=num_samples)  # Binary
      classification for variant presence
    return X, y

# Prepare data
X, y = generate_synthetic_data()
X_train, X_test, y_train, y_test = train_test_split(X, y,
      test_size=0.2, random_state=42)

# Standardize features
```

```python
scaler = StandardScaler()
X_train = scaler.fit_transform(X_train)
X_test = scaler.transform(X_test)

# Define neural network model
def build_nn_model(input_dim):
    model = Sequential([
        Dense(64, input_dim=input_dim, activation='relu'),
        Dropout(0.5),
        Dense(32, activation='relu'),
        Dense(1, activation='sigmoid')  # Output layer for binary
            classification
    ])
    model.compile(optimizer='adam', loss='binary_crossentropy',
        metrics=['accuracy'])
    return model

# Build and train the model
model = build_nn_model(X_train.shape[1])
model.fit(X_train, y_train, epochs=50, batch_size=32,
    validation_split=0.1, verbose=1)

# Evaluate the model
loss, accuracy = model.evaluate(X_test, y_test)
print("Test Accuracy:", accuracy)

# Predict genomic variants
def predict_variants(model, X):
    probabilities = model.predict(X)
    predictions = (probabilities > 0.5).astype(int).flatten()
    return predictions

# Example of predicting on new data
new_data = np.random.rand(10, X_train.shape[1])
new_data_scaled = scaler.transform(new_data)
predictions = predict_variants(model, new_data_scaled)
print("Predicted variants:", predictions)
```

The code defines a neural network model to enhance the detection of genomic variants:

- `generate_synthetic_data` creates synthetic binary-labeled genomic data for model training and evaluation.

- `build_nn_model` constructs a neural network using `Keras` with `Dense` and `Dropout` layers optimized for binary classification.

- The model is trained on standardized synthetic data and its performance evaluated against a test set.

- `predict_variants` utilizes the trained model to predict genomic variants in new samples, demonstrating the potential application of AI in real-world variant detection scenarios.

This comprehensive implementation highlights the role of machine learning in refining genomic variant detection techniques and illustrates a straightforward path to integrating such models into bioinformatics pipelines.

Chapter 11

Deep Learning for Functional Genomics

Below is a Python code snippet that utilizes deep learning frameworks to interpret functional elements in the genome, including data preparation, model implementation using a neural network, and evaluation of model performance.

```python
import numpy as np
import tensorflow as tf
from tensorflow.keras.models import Sequential
from tensorflow.keras.layers import Dense, Dropout
from tensorflow.keras.optimizers import Adam

def preprocess_genome_data(raw_sequences, labels):
    """
    Preprocess raw genomic sequences into a suitable format for
        neural network input.
    :param raw_sequences: List of genomic sequences.
    :param labels: List of labels for functional elements.
    :return: Tuple of feature matrix and labels one-hot encoded.
    """
    # Example encoding: A=1, T=2, C=3, G=4
    seq_length = len(raw_sequences[0])
    feature_matrix = np.zeros((len(raw_sequences), seq_length))

    mapping = {'A': 1, 'T': 2, 'C': 3, 'G': 4}

    for i, seq in enumerate(raw_sequences):
        feature_matrix[i] = [mapping[nuc] for nuc in seq]

    # One-hot encode labels
    labels_encoded = tf.keras.utils.to_categorical(labels)
```

```python
    return feature_matrix, labels_encoded

def build_deep_learning_model(input_shape, num_classes):
    """
    Build a deep learning model for genomic sequence classification.
    :param input_shape: Shape of the input features.
    :param num_classes: Number of target classes.
    :return: Compiled Keras model.
    """
    model = Sequential([
        Dense(128, activation='relu', input_shape=(input_shape,)),
        Dropout(0.2),
        Dense(64, activation='relu'),
        Dropout(0.2),
        Dense(num_classes, activation='softmax')
    ])

    model.compile(optimizer=Adam(),
                  loss='categorical_crossentropy',
                  metrics=['accuracy'])

    return model

def train_and_evaluate(model, x_train, y_train, x_val, y_val,
    ↪  epochs=10, batch_size=32):
    """
    Train the deep learning model and evaluate its performance.
    :param model: Compiled Keras model.
    :param x_train: Training feature data.
    :param y_train: Training labels.
    :param x_val: Validation feature data.
    :param y_val: Validation labels.
    :param epochs: Number of training epochs.
    :param batch_size: Size of training batches.
    :return: Training history and validation accuracy.
    """
    history = model.fit(x_train, y_train,
                        validation_data=(x_val, y_val),
                        epochs=epochs,
                        batch_size=batch_size)

    val_loss, val_accuracy = model.evaluate(x_val, y_val)
    return history, val_accuracy

# Example usage
raw_sequences = ['ATCG', 'GCTA', 'TACG', 'CGTA']
labels = [0, 1, 0, 1]

x, y = preprocess_genome_data(raw_sequences, labels)
x_train, y_train = x[:2], y[:2]  # Example split
x_val, y_val = x[2:], y[2:]
```

```
model = build_deep_learning_model(input_shape=x_train.shape[1],
 ↪  num_classes=2)
history, val_accuracy = train_and_evaluate(model, x_train, y_train,
 ↪  x_val, y_val)

print("Validation Accuracy:", val_accuracy)
```

This code outlines the use of deep learning for interpreting genomic data:

- `preprocess_genome_data` function encodes genomic sequences into numerical formats suitable for machine learning, using one-hot encoding for the labels.

- `build_deep_learning_model` constructs a neural network model using TensorFlow and Keras, designed for classification of functional genomic elements.

- `train_and_evaluate` takes care of model training with a specific number of epochs and batch size, also providing evaluation on validation data to gauge performance.

This provides a baseline framework for applying deep learning to genomic sequence analysis, highlighting preprocessing, model construction, and training phases.

Chapter 12

Graph-based Genomic Data Structures

Below is a Python code snippet that demonstrates the utilization of graph databases and data structures for efficient storage and querying of genomic data using the popular 'networkx' library. This example centers on constructing a graph from genomic interactions and performing typical queries for insights.

```python
import networkx as nx
import matplotlib.pyplot as plt

def create_genomic_graph():
    """
    Creates a directed graph where nodes represent genes or genetic
       markers
    and edges represent interactions or pathways between them.
    :return: A directed graph object.
    """
    G = nx.DiGraph()

    # Add nodes with properties
    genes = ['GeneA', 'GeneB', 'GeneC', 'GeneD']
    for gene in genes:
        G.add_node(gene, expression_level=0.5,
           sample_prop='genomic_marker')

    # Add directed edges between nodes
    interactions = [
        ('GeneA', 'GeneB', {'interaction_type': 'activation'}),
        ('GeneB', 'GeneC', {'interaction_type': 'inhibition'}),
        ('GeneA', 'GeneD', {'interaction_type': 'activation'}),
```

```
            ('GeneC', 'GeneD', {'interaction_type': 'activation'}),
    ]
    G.add_edges_from(interactions)

    return G

def visualize_genomic_graph(G):
    """
    Visualize the genomic graph.
    :param G: The graph object representing genomic data.
    """
    pos = nx.spring_layout(G)
    edge_labels = nx.get_edge_attributes(G, 'interaction_type')
    nx.draw(G, pos, with_labels=True, node_color='lightblue',
    ↪    node_size=2000, arrowsize=20)
    nx.draw_networkx_edge_labels(G, pos, edge_labels=edge_labels)
    plt.title('Genomic Interaction Graph')
    plt.show()

def query_graph_data(G, gene):
    """
    Query the graph to find all nodes directly connected to a given
    ↪    gene.
    :param G: The graph object representing genomic data.
    :param gene: The gene node to query connections from.
    :return: List of connected nodes.
    """
    if gene in G:
        connected_nodes = list(G.successors(gene))
        return connected_nodes
    else:
        return []

# Example usage
genomic_graph = create_genomic_graph()
visualize_genomic_graph(genomic_graph)
result = query_graph_data(genomic_graph, 'GeneA')

print("Genes connected to GeneA:", result)
```

This code involves the following functions to efficiently manage genomic data using graph structures:

- `create_genomic_graph` constructs a directed genetic interaction graph, node properties representing gene expression levels.

- `visualize_genomic_graph` provides a visual representation of the network, emphasizing connectivity and interaction types.

- `query_graph_data` retrieves and lists genes that are directly influenced by a specified gene, facilitating the analysis of genetic pathways.

Such a graph model provides a robust framework for exploring and interpreting genomic relationships, enhancing storage and data manipulation efficiency.

Chapter 13

Phylogenetic Tree Construction Algorithms

Below is a Python code snippet illustrating advanced algorithms for constructing and analyzing phylogenetic trees from genetic data.

```
from Bio import Phylo
from Bio.Phylo.TreeConstruction import DistanceCalculator
from Bio.Phylo.TreeConstruction import DistanceTreeConstructor
from Bio.Align import MultipleSeqAlignment
from Bio.Seq import Seq
from Bio.SeqRecord import SeqRecord

# Example sequences
sequences = [
    SeqRecord(Seq("ATCGTAC"), id="Alpha"),
    SeqRecord(Seq("ATCGTGC"), id="Beta"),
    SeqRecord(Seq("ATGGTAC"), id="Gamma"),
    SeqRecord(Seq("ATCGTCC"), id="Delta")
]

# Creating a Multiple Sequence Alignment
alignment = MultipleSeqAlignment(sequences)

# Calculate the distance matrix
calculator = DistanceCalculator('identity')
dm = calculator.get_distance(alignment)

# Constructing the Tree using UPGMA algorithm
constructor = DistanceTreeConstructor(calculator)
```

```
tree = constructor.upgma(dm)

# Writing the tree to the terminal in a simple text format
Phylo.draw_ascii(tree)

# To export the tree in Newick format:
Phylo.write(tree, "tree.newick", "newick")

# Function to count the internal nodes of a phylogenetic tree
def count_internal_nodes(tree):
    '''
    Count the internal nodes in a phylogenetic tree.
    :param tree: A Phylo Tree object.
    :return: Number of internal nodes.
    '''
    return len(tree.get_nonterminals())

# Example of counting internal nodes
internal_nodes_count = count_internal_nodes(tree)
print("Number of internal nodes:", internal_nodes_count)
```

This code demonstrates core components required to construct and analyze phylogenetic trees from genetic sequences:

- `DistanceCalculator` computes the pairwise distance matrix between sequences.

- `DistanceTreeConstructor` is used to form trees from the distance matrix, utilizing methods like UPGMA.

- The example creates a simple set of sequences, aligns them, calculates a distance matrix, and constructs a tree.

- `count_internal_nodes` is a function to assess tree complexity by counting its internal nodes.

The structure also provides an approach to export trees in standardized formats and visualize them, aiding further phylogenetic analysis.

Chapter 14

Maximum Likelihood Estimation in Phylogenetics

Below is a Python code snippet that implements Maximum Likelihood Estimation (MLE) methods for estimating evolutionary relationships among species using phylogenetic trees and substitution models. We use the `scipy` library for optimization, and the `Bio.Phylo` module from Biopython for handling phylogenetic trees.

```
import numpy as np
from Bio import Phylo
from scipy.optimize import minimize
from numpy.linalg import matrix_power

def jc69_matrix(alpha, time):
    '''
    Constructs the Jukes-Cantor substitution matrix.
    :param alpha: Substitution rate.
    :param time: Evolutionary time.
    :return: Substitution matrix.
    '''
    p = 1 - 3/4 * np.exp(-4/3 * alpha * time)
    q = 1/4 * np.exp(-4/3 * alpha * time)
    return np.array([[p, q, q, q],
                     [q, p, q, q],
                     [q, q, p, q],
                     [q, q, q, p]])

def likelihood_function(branch_lengths, tree, data, alpha):
```

```
'''
Calculates the likelihood of observing data given branch lengths
↪ and model parameters.
:param branch_lengths: An array of branch lengths.
:param tree: Phylogenetic tree object.
:param data: Observed sequences.
:param alpha: Rate of substitution.
:return: Likelihood value.
'''
likelihood = 0.0
for i, clade in enumerate(tree.get_terminals()):
    P = jc69_matrix(alpha, branch_lengths[i])
    likelihood += np.sum(data[clade.name] * np.log(P))
return -likelihood   # We return negative likelihood for
↪ minimization

def estimate_branch_lengths(tree, data, alpha):
    '''
    Estimates the most likely branch lengths using MLE.
    :param tree: Phylogenetic tree object.
    :param data: Observed sequence data.
    :param alpha: Rate of substitution.
    :return: Optimized branch lengths.
    '''
    initial_guess = np.ones(len(tree.get_terminals()))   # Initial
    ↪ lengths set to 1
    result = minimize(likelihood_function, initial_guess,
    ↪ args=(tree, data, alpha), bounds=[(0.01,
    ↪ 2.0)]*len(tree.get_terminals()))
    return result.x

# Sample Phylogenetic Tree for demonstration (Newick format)
tree = Phylo.read('example_tree.newick', 'newick')

# Example observed sequence data (Probabilities for ACGT)
data = {
    'Species1': np.array([0.9, 0.05, 0.03, 0.02]),
    'Species2': np.array([0.88, 0.07, 0.03, 0.02]),
    'Species3': np.array([0.85, 0.1, 0.03, 0.02]),
    'Species4': np.array([0.87, 0.06, 0.05, 0.02])
}

# Substitution rate
alpha = 0.5

optimized_branch_lengths = estimate_branch_lengths(tree, data,
↪ alpha)
print("Optimized Branch Lengths:", optimized_branch_lengths)
```

This code defines several key functions necessary for implementing MLE methods in phylogenetic analysis:

- `jc69_matrix` constructs the Jukes-Cantor substitution matrix for nucleotide changes.

- `likelihood_function` computes the likelihood of the observed sequence data given a set of branch lengths and a substitution rate.

- `estimate_branch_lengths` uses optimization techniques to estimate the branch lengths by maximizing the likelihood function.

To use this code, provide a phylogenetic tree in Newick format and observed sequence data. The script estimates the branch lengths using the Jukes-Cantor model, allowing for an interpretation of evolutionary relationships. Make sure to adapt the input file name and sequence data for your specific use case.

Chapter 15

Bayesian Inference in Evolutionary Biology

Below is a Python code snippet that encompasses the core computational elements of applying Bayesian models to infer evolutionary parameters and phylogenies, including the construction of phylogenetic trees, handling of evolutionary models, and Bayesian inference using sampling techniques.

```python
import numpy as np
from scipy.stats import norm
import matplotlib.pyplot as plt

def simulate_evolutionary_data(num_species, num_sites):
    """
    Simulates evolutionary data for a given number of species and
      sites.
    :param num_species: Number of species.
    :param num_sites: Number of sequence sites.
    :return: Simulated sequence data.
    """
    return np.random.choice(['A', 'C', 'G', 'T'], size=(num_species,
      num_sites))

def calculate_likelihood(sequence_data, tree_topology,
  model_parameters):
    """
    Calculate the likelihood of the sequence data given a tree
      topology and model parameters.
    :param sequence_data: Simulated or observed sequence data.
    :param tree_topology: Array representing the branch structure of
      the tree.
    :param model_parameters: Parameters of the evolutionary model.
```

```
    :return: Likelihood value.
    """
    likelihood = 1.0
    for site in sequence_data.T:
        # Simplified model: assume a uniform prior
        site_likelihood = np.prod([model_parameters.get(base, 0.25)
         ↪  for base in site])
        likelihood *= site_likelihood
    return likelihood

def bayesian_inference(sequence_data, num_iterations, priors):
    """
    Perform Bayesian inference over phylogenetic tree space.
    :param sequence_data: Sequence data to analyse.
    :param num_iterations: Number of iterations for MCMC.
    :param priors: Prior distributions on model parameters.
    :return: Samples of trees and model parameters.
    """
    sampled_trees = []
    current_tree = np.random.rand(sequence_data.shape[0] - 1, 2)  #
     ↪  Random initial tree

    for _ in range(num_iterations):
        proposed_tree = current_tree + np.random.normal(scale=0.1,
         ↪  size=current_tree.shape)
        model_parameters = {base: np.random.rand() for base in ['A',
         ↪  'C', 'G', 'T']}

        # Calculate likelihoods
        current_likelihood = calculate_likelihood(sequence_data,
         ↪  current_tree, model_parameters)
        proposed_likelihood = calculate_likelihood(sequence_data,
         ↪  proposed_tree, model_parameters)

        # Calculate acceptance ratio assuming uniform prior
        acceptance_ratio = proposed_likelihood / current_likelihood

        # Metropolis criteria
        if np.random.rand() < acceptance_ratio:
            current_tree = proposed_tree
            sampled_trees.append(current_tree)

    return sampled_trees

def plot_sampled_trees(sampled_trees):
    """
    Plot a trace plot of sampled tree likelihoods.
    :param sampled_trees: List of trees sampled during Bayesian
     ↪  inference.
    """
```

```
    likelihoods =
 ↪   [calculate_likelihood(simulate_evolutionary_data(10, 50),
 ↪   tree, {'A':0.25, 'C':0.25, 'G':0.25, 'T':0.25}) for tree in
 ↪   sampled_trees]
    plt.plot(likelihoods)
    plt.xlabel('Iteration')
    plt.ylabel('Likelihood')
    plt.title('Trace plot of sampled tree likelihoods')
    plt.show()

# Example usage
sequence_data = simulate_evolutionary_data(4, 10)
sampled_trees = bayesian_inference(sequence_data, 1000,
 ↪   {'frequencies': norm(0.25, 0.1)})

plot_sampled_trees(sampled_trees)
```

This code defines key functions necessary for applying Bayesian models to infer evolutionary parameters and phylogenies:

- `simulate_evolutionary_data` generates simulated sequence data for a set number of species and sites.

- `calculate_likelihood` computes the likelihood of sequence data given a tree topology and model parameters.

- `bayesian_inference` performs Bayesian inference using a Metropolis algorithm to sample possible phylogenetic trees.

- `plot_sampled_trees` visualizes the trace plot of the likelihoods of sampled trees to assess convergence and distribution.

The final block of code demonstrates an entire procedure from data simulation to Bayesian inference and visualization, enabling comprehensive understanding and application of Bayesian phylogenetic analysis.

Chapter 16

Coalescent Theory Algorithms

Below is a Python code snippet that illustrates the implementation of a coalescent theory simulation, which models genetic variation over time by considering lineages descending from a common ancestor and their branching patterns.

```python
import numpy as np
import matplotlib.pyplot as plt


def simulate_coalescent(n, population_size, generations):
    '''
    Simulates coalescent process for a given number of samples.
    :param n: Number of samples (e.g., individuals' genes).
    :param population_size: Constant effective population size.
    :param generations: Number of generations to simulate.
    :return: List of coalescence times between lineages.
    '''
    lineage_count = n
    coalescence_times = []
    current_generation = 0

    # While more than one lineage remains
    while lineage_count > 1:
        # Exponentially distributed waiting time with mean 2N / (k *
        #   (k - 1))
        wait_time = np.random.exponential(2 * population_size /
            (lineage_count * (lineage_count - 1)))
        current_generation += wait_time
        coalescence_times.append(current_generation)
```

```python
        # One coalescence event reduces the number of lineages by 1
        lineage_count -= 1

    return coalescence_times


def plot_coalescence(tree_times, label):
    '''
    Plots the coalescence time for a given simulation.
    :param tree_times: Coalescence times to plot.
    :param label: Description label for the plot.
    '''
    plt.step(range(len(tree_times) + 1), [0] + tree_times,
     label=label, where='post')
    plt.xlabel('Number of ancestral lineages')
    plt.ylabel('Time')
    plt.title('Coalescence Simulation')
    plt.legend()


# Example parameters
n_samples = 10
pop_size = 500
max_generations = 1000

# Perform simulation
coalescence_times = simulate_coalescent(n_samples, pop_size,
     max_generations)

# Plotting the result
plot_coalescence(coalescence_times, f'{n_samples} Samples,
     Population Size: {pop_size}')
plt.show()
```

This code encapsulates the key aspects of a coalescent theory simulation, which include:

- `simulate_coalescent` function calculates the coalescence times for a number of lineages, assuming a constant population size and given number of generations.

- `plot_coalescence` helps in visualizing the tree-like structure showing coalescence times and intervals between lineages over generations.

The example parameters illustrate a scenario with 10 samples in a constant population size of 500, creating a visual representation of the coalescent tree generated through this simulation.

Chapter 17

Genome-Wide Association Study (GWAS) Algorithms

Below is a Python code snippet that encompasses the core computational elements for identifying associations between genetic variants and traits, including the loading of genetic data, preprocessing, applying statistical tests, and utilizing machine learning models for association discovery.

```python
import pandas as pd
import numpy as np
from sklearn.model_selection import train_test_split
from sklearn.ensemble import RandomForestClassifier
from sklearn.metrics import classification_report, roc_auc_score
from scipy import stats

def load_genetic_data(file_path):
    '''
    Load genetic data from a file (CSV format for simplicity).
    :param file_path: Path to the CSV file containing genetic data.
    :return: DataFrame containing genetic information.
    '''
    return pd.read_csv(file_path)

def preprocess_data(genetic_data, trait_data):
    '''
    Preprocess the data for analysis.
    :param genetic_data: DataFrame containing the genetic variants.
    :param trait_data: Series containing the trait of interest.
```

```python
    :return: Preprocessed features and labels.
    '''
    # Convert categorical variables to one-hot encoding
    genetic_data_encoded = pd.get_dummies(genetic_data)
    # Ensure the trait data is binary for classification
    trait_data_binary = (trait_data >
    ↪   trait_data.median()).astype(int)
    return genetic_data_encoded, trait_data_binary

def perform_statistical_tests(genetic_data, trait_data):
    '''
    Perform statistical tests to find significant genetic variants.
    :param genetic_data: DataFrame of genetic variants.
    :param trait_data: Series of the trait data.
    :return: List of significant variants based on p-values.
    '''
    significant_variants = []
    for variant in genetic_data.columns:
        stat, p_value = stats.ttest_ind(genetic_data[variant],
        ↪   trait_data)
        if p_value < 0.05:  # Using 0.05 as a significance threshold
            significant_variants.append(variant)
    return significant_variants

def train_ml_model(X, y):
    '''
    Train a machine learning model to predict the genetic
    ↪   association.
    :param X: Feature matrix.
    :param y: Target vector.
    :return: Trained model.
    '''
    X_train, X_test, y_train, y_test = train_test_split(X, y,
    ↪   test_size=0.2, random_state=42)
    model = RandomForestClassifier(n_estimators=100,
    ↪   random_state=42)
    model.fit(X_train, y_train)
    y_pred = model.predict(X_test)
    print("Classification Report:\n", classification_report(y_test,
    ↪   y_pred))
    print("ROC AUC Score:", roc_auc_score(y_test,
    ↪   model.predict_proba(X_test)[:,1]))
    return model

# Example workflow execution
file_path = 'genetic_data.csv'
genetic_data = load_genetic_data(file_path)
trait_data = genetic_data['trait']
genetic_data_features = genetic_data.drop(columns=['trait'])

# Preprocess and find significant genetic variants
X, y = preprocess_data(genetic_data_features, trait_data)
```

```
significant_vars = perform_statistical_tests(genetic_data_features,
 ↪   trait_data)
print("Significant Variants:", significant_vars)

# Train a machine learning model on the significant variants
X_significant = X[significant_vars]
ml_model = train_ml_model(X_significant, y)
```

This Python code defines several functions necessary for the implementation of computational methods to identify associations between genetic variants and traits:

- `load_genetic_data` function loads the genetic data from a CSV file for analysis.

- `preprocess_data` converts categorical genetic data to numeric forms and binarizes the trait data.

- `perform_statistical_tests` applies statistical tests to identify significant genetic variants.

- `train_ml_model` trains a machine learning model using significant genetic variants to predict trait association.

The script demonstrates the steps of preprocessing the data, performing initial statistical analysis, and utilizing machine learning for further trait association predictions.

Chapter 18

Linkage Disequilibrium Mapping Techniques

Below is a Python code snippet that demonstrates how to implement an algorithm for mapping genes based on linkage disequilibrium in populations.

```python
import numpy as np
from scipy.stats import chi2
import matplotlib.pyplot as plt

def calculate_ld_r2(gene1, gene2):
    '''
    Calculate the linkage disequilibrium measure r^2 between two
    ↪ genes.
    :param gene1: Allele frequencies for gene1.
    :param gene2: Allele frequencies for gene2.
    :return: r^2 value.
    '''
    p_A1, p_a1 = gene1
    p_B1, p_b1 = gene2
    p_AB1 = p_A1 * p_B1   # Assuming random association initially

    D = p_AB1 - (p_A1 * p_B1)
    r_squared = (D ** 2) / (p_A1 * p_a1 * p_B1 * p_b1)
    return r_squared

def test_for_ld(gene1, gene2, n):
    '''
    Perform a chi-square test for independence to detect linkage
    ↪ disequilibrium.
    :param gene1: Allele frequencies for gene1.
    :param gene2: Allele frequencies for gene2.
```

```python
    :param n: Sample size.
    :return: chi-square statistic and p-value.
    '''
    r_squared = calculate_ld_r2(gene1, gene2)
    chi_sq = n * r_squared
    p_value = chi2.sf(chi_sq, 1)
    return chi_sq, p_value

def generate_random_allele_frequencies():
    '''
    Generate random allele frequencies ensuring they sum to 1.
    :return: Tuple of (allele frequency, complementary frequency).
    '''
    freq = np.random.rand()
    return freq, 1 - freq

def main():
    '''
    Main function to simulate linkage disequilibrium mapping for
    ↪ multiple genes.
    '''
    num_genes = 100  # Number of genes to simulate
    sample_size = 1000

    gene_ld_matrix = np.zeros((num_genes, num_genes))  # Matrix to
    ↪ store r^2 values

    for i in range(num_genes):
        for j in range(i+1, num_genes):
            gene1 = generate_random_allele_frequencies()
            gene2 = generate_random_allele_frequencies()
            r_squared = calculate_ld_r2(gene1, gene2)
            gene_ld_matrix[i, j] = r_squared
            gene_ld_matrix[j, i] = r_squared

    plt.imshow(gene_ld_matrix, cmap='hot', interpolation='nearest')
    plt.title("Linkage Disequilibrium r^2 Heatmap")
    plt.colorbar()
    plt.show()

if __name__ == "__main__":
    main()
```

This Python code performs the following tasks related to mapping genes based on linkage disequilibrium:

- `calculate_ld_r2` function computes the linkage disequilibrium measure r^2 between two genes.

- `test_for_ld` computes the chi-square statistic to test for statistical significance of LD.

- `generate_random_allele_frequencies` generates random allele frequencies summing to 1, demonstrating random initial conditions.

- `main` function simulates the LD mapping, calculating r^2 values for a number of genes and visualizing them in a heatmap.

The results provide visual insights into the degree of linkage disequilibrium among simulated genes.

Chapter 19

Population Genetics Simulation Algorithms

Below is a Python code snippet that simulates genetic variation in populations using computational models. The code implements a basic Wright-Fisher model to simulate allele frequency changes over generations under the influence of genetic drift.

```
import numpy as np
import matplotlib.pyplot as plt

def wright_fisher_simulation(p0, population_size, generations,
    n_simulations):
    '''
    Simulate genetic variation over generations using the
        Wright-Fisher model.
    :param p0: Initial allele frequency.
    :param population_size: Number of individuals in the population.
    :param generations: Number of generations to simulate.
    :param n_simulations: Number of simulations to run.
    :return: List of allele frequencies over generations for each
        simulation.
    '''
    # Store results for each simulation
    simulation_results = []

    for sim in range(n_simulations):
        p = p0
        frequencies = [p]

        for gen in range(generations):
            # Binomial sampling to simulate reproduction
```

```
            p = np.random.binomial(population_size, p) /
            ↪  population_size
            frequencies.append(p)

        simulation_results.append(frequencies)

    return simulation_results

# Parameters for the simulation
initial_frequency = 0.5
population_size = 100
generations = 50
n_simulations = 5

# Run the Wright-Fisher simulations
simulation_data = wright_fisher_simulation(initial_frequency,
↪  population_size, generations, n_simulations)

# Plot the results
for sim_index in range(n_simulations):
    plt.plot(simulation_data[sim_index], label=f'Simulation
    ↪  {sim_index + 1}')

plt.xlabel('Generations')
plt.ylabel('Allele Frequency')
plt.title('Wright-Fisher Model Simulations of Allele Frequency')
plt.legend()
plt.grid(True)
plt.show()
```

This code provides an implementation of the Wright-Fisher model, commonly used in population genetics to understand genetic drift:

- `wright_fisher_simulation` simulates allele frequency changes using a binomial sampling approach to capture genetic drift across generations. The function takes initial allele frequency, population size, the number of generations, and the number of simulations to run as inputs and returns the simulated allele frequencies.

- The parameters defined include initial allele frequency, population size, the number of generations to simulate, and the number of simulations to perform.

- The outputs are visualized using a simple line plot where each line represents a different simulation of allele frequency changes over time.

The plot generated helps to visually understand how stochastic genetic drift can lead to changes in allele frequencies over time in small populations.

Chapter 20

Metabolic Pathway Modeling and Simulation

Below is a Python code snippet that encompasses the core computational elements for modeling and simulating metabolic pathways in silico, including setup of the metabolic network, flux balance analysis (FBA), and visualization of the results.

```
import numpy as np
import matplotlib.pyplot as plt
from scipy.optimize import linprog

class MetabolicNetwork:
    def __init__(self, S, c, b):
        """
        Initialize the metabolic network.

        :param S: Stoichiometric matrix.
        :param c: Objective coefficients for linear programming.
        :param b: Flux availability vector.
        """
        self.S = S
        self.c = c
        self.b = b

    def perform_fba(self, method='highs'):
        """
        Perform Flux Balance Analysis (FBA) by linear programming.
```

```
            :param method: Optimization method ('highs', 'simplex',
            ↪  etc.).
            :return: Flux distribution.
            """
            res = linprog(c=self.c, A_eq=self.S, b_eq=self.b,
            ↪  method=method)
            if res.success:
                return res.x
            else:
                raise Exception("Optimization failed:", res.message)

def build_example_network():
    """
    Set up an example metabolic network.

    :return: Metabolic network instance.
    """
    # Example stoichiometric matrix (S) and vectors (c, b)
    S = np.array([[1, -1, 0], [0, 1, -1]])  # Example stoichiometry
    c = np.array([-1, 0, 0])                 # Maximize first flux
    b = np.array([0, 0])                     # Balance equations are
    ↪  zero

    return MetabolicNetwork(S, c, b)

def plot_flux_distribution(fluxes):
    """
    Plot a simple flux distribution.

    :param fluxes: Array of flux values obtained from FBA.
    """
    rxns = ['R1', 'R2', 'R3']
    plt.bar(rxns, fluxes)
    plt.title("Flux Distribution")
    plt.xlabel("Reactions")
    plt.ylabel("Flux Value")
    plt.show()

# Example usage
network = build_example_network()
fluxes = network.perform_fba()
print("Optimal flux distribution:", fluxes)
plot_flux_distribution(fluxes)
```

This code defines several key components necessary for modeling and simulating metabolic pathways:

- `MetabolicNetwork` class encapsulates a stoichiometric matrix and related vectors for metabolic analysis.

- `perform_fba` method uses linear programming to compute flux distributions, central to simulating metabolic behaviors.

- `build_example_network` function sets up a simple example network to perform FBA on, representing a conceptual metabolic model.

- `plot_flux_distribution` function visualizes the results as a bar chart, helping in analyzing distribution over reactions.

This snippet introduces a structure for simulating metabolic systems by setting objectives and balance conditions, where actual biological systems can replace the provided mock setup.

Chapter 21

Flux Balance Analysis in Metabolic Engineering

Below is a Python code snippet that highlights the core computational methods for predicting metabolic fluxes in metabolic networks using Flux Balance Analysis (FBA) and includes the setup of the metabolic model, simulation of the flux distribution, and interpretation of the results.

```
import numpy as np
from scipy.optimize import linprog

def create_metabolic_network():
    """
    Create a simple metabolic network represented by a
    ↪ stoichiometric matrix.
    :return: Stoichiometric matrix and objective vector.
    """
    # Example stoichiometric matrix (rows: metabolites, columns:
    ↪ reactions)
    S = np.array([
        [-1, -1,  0,  1],    # Metabolite A
        [ 1,  0, -1,  0],    # Metabolite B
        [ 0,  1,  1, -1]     # Metabolite C
    ])

    # Objective vector (maximize production of metabolite C)
    c = np.array([0, 0, 0, 1])
```

```python
    return S, c

def perform_flux_balance_analysis(S, c):
    """
    Perform Flux Balance Analysis (FBA) on the given metabolic
    ↪ network.
    :param S: Stoichiometric matrix.
    :param c: Objective vector.
    :return: Optimal flux distribution.
    """
    num_reactions = S.shape[1]

    # Bounds for each reaction flux, assuming irreversible reactions
    bounds = [(0, 1000) for _ in range(num_reactions)]

    # Solve linear programming problem to maximize the objective
    res = linprog(c=-c, A_eq=S, b_eq=np.zeros(S.shape[0]),
    ↪ bounds=bounds, method='highs')

    if res.success:
        return res.x
    else:
        raise ValueError("FBA optimization failed.")

def main():
    """
    Main function to set up and solve the FBA problem.
    """
    S, c = create_metabolic_network()

    try:
        optimal_flux_distribution = perform_flux_balance_analysis(S,
        ↪ c)
        print("Optimal Flux Distribution:",
        ↪ optimal_flux_distribution)
    except ValueError as e:
        print(str(e))

if __name__ == "__main__":
    main()
```

This code defines essential functions for implementing Flux Balance Analysis (FBA) for metabolic networks, which includes:

- `create_metabolic_network` sets up a simple stoichiometric matrix representing the metabolic network, along with the objective function vector.

- `perform_flux_balance_analysis` executes the FBA by solving a linear programming problem to find the optimal flux distribution that maximizes the objective.

- The main function, `main`, orchestrates the setup and solution of the FBA problem, handling success and failure outcomes.

The final segment of the code demonstrates running the analysis on a sample network, outputting the optimal flux distribution if successful.

Chapter 22

Constraint-Based Reconstruction and Analysis (COBRA) Methods

Below is a Python code snippet that encompasses the core computational elements of reconstructing and analyzing genome-scale metabolic models using the COBRApy package, focusing on model setup, simulation, and analysis.

```
import cobra
from cobra import Model, Reaction, Metabolite

def create_metabolic_model():
    '''
    Create a simple metabolic model using COBRApy.
    :return: A COBRA model object.
    '''
    model = Model("example_model")

    # Create and add metabolites
    metabolite_a = Metabolite('A', formula='C3H8O3',
        name='Metabolite A', compartment='c')
    metabolite_b = Metabolite('B', formula='C3H4O3',
        name='Metabolite B', compartment='c')
    metabolite_c = Metabolite('C', formula='C3H6O3',
        name='Metabolite C', compartment='c')
```

```python
    # Create and add reactions
    reaction1 = Reaction('reaction1')
    reaction1.name = 'Reaction 1'
    reaction1.lower_bound = 0
    reaction1.upper_bound = 1000
    reaction1.add_metabolites({metabolite_a: -1.0, metabolite_b:
    ↪   1.0})

    reaction2 = Reaction('reaction2')
    reaction2.name = 'Reaction 2'
    reaction2.lower_bound = 0
    reaction2.upper_bound = 1000
    reaction2.add_metabolites({metabolite_b: -1.0, metabolite_c:
    ↪   1.0})

    model.add_reactions([reaction1, reaction2])

    # Set the objective
    model.objective = 'reaction2'

    return model

def simulate_fba(model):
    '''
    Perform Flux Balance Analysis on the model.
    :param model: The COBRA model object.
    :return: The solution of the FBA.
    '''
    solution = model.optimize()
    return solution

def analyze_solution(solution):
    '''
    Analyze the solution of the FBA to extract and print flux
    ↪   values.
    :param solution: The solution of the FBA.
    '''
    print("Objective value:", solution.objective_value)
    print("Fluxes:")
    for reaction in solution.fluxes.index:
        print(f"{reaction}: {solution.fluxes[reaction]}")

# Main execution
metabolic_model = create_metabolic_model()
fba_solution = simulate_fba(metabolic_model)
analyze_solution(fba_solution)
```

This code defines several key functions necessary for setting up and analyzing a genome-scale metabolic model using COBRApy:

- `create_metabolic_model` function constructs a simple COBRA metabolic model by defining metabolites and reactions, and adding them to the model.

- `simulate_fba` performs Flux Balance Analysis (FBA) on the model to predict the flow of metabolites through the metabolic reactions.

- `analyze_solution` extracts the objective value and individual reaction fluxes from the FBA solution and prints them for analysis.

The final block of code demonstrates the creation, simulation, and analysis of the metabolic model, providing insights into metabolic flux distributions.

Chapter 23

Synthetic Gene Circuit Design Algorithms

Below is a Python code snippet dedicated to the design of synthetic gene circuits. This snippet includes the simulation of gene circuit behavior, optimization of gene expression, and visualization of the synthetic gene network using core computational principles.

```
import numpy as np
import matplotlib.pyplot as plt
from scipy.integrate import odeint

def gene_circuit_odes(y, t, params):
    '''
    Define the ODEs for the synthetic gene circuit.
    :param y: Current state of the circuit.
    :param t: Time.
    :param params: Parameters of the model.
    :return: Derivatives.
    '''
    # Unpack state variables and parameters
    A, B = y
    alpha, beta, K, n = params

    # ODEs
    dA_dt = alpha / (1 + (B/K)**n) - A
    dB_dt = beta / (1 + (A/K)**n) - B

    return [dA_dt, dB_dt]

def simulate_gene_circuit(initial_conditions, params, t_span):
    '''
    Simulate the gene circuit dynamics over time.
```

```
    :param initial_conditions: Initial concentrations of components.
    :param params: Parameters of the model.
    :param t_span: Time span for the simulation.
    :return: Time array and concentrations over time.
    '''
    result = odeint(gene_circuit_odes, initial_conditions, t_span,
    ↪   args=(params,))
    return t_span, result

def plot_gene_circuit(t, concentrations):
    '''
    Plot the concentrations of gene circuit components over time.
    :param t: Time array.
    :param concentrations: Concentrations array.
    '''
    plt.plot(t, concentrations[:, 0], label='Protein A')
    plt.plot(t, concentrations[:, 1], label='Protein B')
    plt.xlabel('Time')
    plt.ylabel('Concentration')
    plt.title('Synthetic Gene Circuit Dynamics')
    plt.legend()
    plt.show()

# Define parameters and initial conditions
params = (1.5, 1.0, 0.5, 2)   # (alpha, beta, K, n)
initial_conditions = [0, 0]
t_span = np.linspace(0, 50, 100)

# Simulate and visualize the gene circuit
t, concentrations = simulate_gene_circuit(initial_conditions,
↪   params, t_span)
plot_gene_circuit(t, concentrations)
```

This code provides a comprehensive approach to modeling and designing synthetic gene circuits with desired behaviors:

- `gene_circuit_odes` defines the system of ordinary differential equations (ODEs) representing the gene circuit.

- `simulate_gene_circuit` uses numerical integration to simulate the changes in concentrations of gene products over time.

- `plot_gene_circuit` facilitates the visualization of how protein levels evolve within the synthetic circuit.

These components work together to model the dynamic behavior of synthetic gene circuits, important for understanding and designing circuits with specific functions and responses.

Chapter 24

Boolean Network Modeling of Gene Regulatory Networks

Below is a Python code snippet that demonstrates the modeling and analysis of gene regulatory interactions using Boolean networks. It includes the simulation of a Boolean network, its dynamics over iterations, and the visualization of gene states.

```
import numpy as np
import networkx as nx
import matplotlib.pyplot as plt

class BooleanNetwork:
    def __init__(self, nodes, edges, initial_states):
        """
        Initialize the Boolean network with nodes, edges, and
        ↪ initial states.
        :param nodes: List of node names (genes).
        :param edges: List of tuples representing edges (gene
        ↪ interactions).
        :param initial_states: Dictionary of initial states for each
        ↪ node.
        """
        self.graph = nx.DiGraph()
        self.graph.add_nodes_from(nodes)
        self.graph.add_edges_from(edges)
        self.states = initial_states

    def update_state(self, update_rules):
        """
```

```
            Update the state of each node based on the provided update
         ↪   rules.
            :param update_rules: Dictionary with nodes as keys and a
         ↪   function to update state based on inputs.
            """
            new_states = {}
            for node in self.graph.nodes:
                inputs = [self.states[neighbor] for neighbor in
                 ↪   self.graph.predecessors(node)]
                new_states[node] = update_rules[node](*inputs)
            self.states = new_states

        def simulate(self, steps, update_rules):
            """
            Simulate the network over a specified number of steps.
            :param steps: Number of simulation steps.
            :param update_rules: Dictionary with nodes as keys and
         ↪   update functions.
            :return: List of state dictionaries over time.
            """
            history = []
            for _ in range(steps):
                history.append(self.states.copy())
                self.update_state(update_rules)
            return history

        def plot_network(self):
            """
            Plot the current state of the Boolean network.
            """
            plt.figure(figsize=(8, 6))
            node_colors = ['red' if self.states[node] else 'blue' for
         ↪   node in self.graph.nodes]
            nx.draw(self.graph, with_labels=True,
         ↪   node_color=node_colors, node_size=800, font_size=10)
            plt.show()

# Example usage

# Define nodes (genes) and edges (interactions)
nodes = ['A', 'B', 'C']
edges = [('A', 'B'), ('B', 'C'), ('C', 'A')]

# Initial states of the genes
initial_states = {'A': 1, 'B': 0, 'C': 0}

# Define update rules for each node
update_rules = {
    'A': lambda B: B,        # A depends on B
    'B': lambda A: A,        # B depends on A
    'C': lambda B: not B     # C depends inversely on B
}
```

```
# Initialize the Boolean Network
network = BooleanNetwork(nodes, edges, initial_states)

# Simulate the network for 10 steps
history = network.simulate(steps=10, update_rules=update_rules)

# Plot the network at the final state
network.plot_network()

# Print out the history of states
for t, states in enumerate(history):
    print(f"Time {t}: {states}")
```

This code defines a 'BooleanNetwork' class to simulate gene regulatory interactions using Boolean logic:

- `BooleanNetwork.__init__`: Initializes the network with genes (nodes), interactions (edges), and initial states.

- `update_state`: Updates each node's state based on incoming signals, defined by update rules.

- `simulate`: Runs the simulation for a predefined number of time steps, recording the states at each step.

- `plot_network`: Visualizes the gene network, using color to indicate gene states (e.g., red for active, blue for inactive).

The example usage demonstrates setting up a simple network with three genes, defining their interactions, and running the simulation to explore the dynamic behavior over time. The network's state is visualized, and the evolution of states is printed to provide insight into gene regulatory dynamics.

Chapter 25

Stochastic Simulations of Cellular Processes

Below is a Python code snippet that details the implementation of stochastic algorithms to simulate random events in cellular processes. This includes the setup of cellular states, stochastic event simulation, and statistical analysis of simulation outcomes.

```
import numpy as np
import matplotlib.pyplot as plt

def initialize_states(n):
    '''
    Initialize the state of each cell in the cellular system.
    :param n: Number of cells.
    :return: Initial states as an array.
    '''
    # Randomly initialize each cell to 0 (inactive) or 1 (active)
    return np.random.choice([0, 1], size=n)

def simulate_event(states, rate):
    '''
    Simulate the random occurrence of events in cellular states.
    :param states: Current state of the system.
    :param rate: Transition rate for state changes.
    :return: Updated states after event simulation.
    '''
    # Compute the probability of state change for each cell
    probability_of_change = rate * np.random.random(states.shape)
    # Determine which states change
    changes = probability_of_change > 0.5
    states = np.where(changes, 1 - states, states)
    return states
```

```python
def simulate_process(n_cells, n_steps, transition_rate):
    '''
    Run the stochastic simulation over a series of time steps.
    :param n_cells: Number of cells.
    :param n_steps: Number of simulation steps.
    :param transition_rate: Rate at which cells change state.
    :return: Time series of active states.
    '''
    states = initialize_states(n_cells)
    activity_over_time = [np.sum(states)]

    for step in range(n_steps):
        states = simulate_event(states, transition_rate)
        activity_over_time.append(np.sum(states))

    return activity_over_time

def plot_activity(activity_over_time):
    '''
    Plot the number of active cells over time.
    :param activity_over_time: Time series data of active cell
    ↪ counts.
    '''
    plt.figure(figsize=(10, 6))
    plt.plot(activity_over_time)
    plt.title('Stochastic Simulation of Cellular Activity')
    plt.xlabel('Time Steps')
    plt.ylabel('Number of Active Cells')
    plt.show()

# Parameters for the simulation
num_cells = 100
num_steps = 200
transition_rate = 0.1

# Run the simulation
activity_data = simulate_process(num_cells, num_steps,
↪ transition_rate)

# Plot the results
plot_activity(activity_data)
```

This code illustrates several core components crucial for simulating cellular processes using stochastic algorithms:

- `initialize_states` function sets up the initial states of cells in a binary manner (active or inactive).

- `simulate_event` performs the stochastic simulation of state transitions based on a given transition rate.

- `simulate_process` encompasses the iterative simulation over defined time steps, accounting for the number of active changes over time.

- `plot_activity` visualizes the stochastic variation in active cell states throughout the simulation.

The final execution block conducts these simulations utilizing predefined parameters and produces a graphical depiction of the stochastic dynamics in the cell populations.

Chapter 26

Agent-Based Modeling in Systems Biology

Below is a Python code snippet that encompasses the core elements for programming an agent-based model to simulate interactions within biological systems. This example demonstrates how agents can interact in a simulated environment representing biological entities.

```
import random
import matplotlib.pyplot as plt

class Agent:
    '''
    Represents an individual agent in the simulation.
    Each agent has a position and can move around in a grid.
    '''
    def __init__(self, id, x, y, grid_size):
        self.id = id
        self.x = x
        self.y = y
        self.grid_size = grid_size

    def move(self):
        '''Move the agent to a new position within the grid.'''
        dx = random.choice([-1, 0, 1])
        dy = random.choice([-1, 0, 1])
        new_x = (self.x + dx) % self.grid_size
        new_y = (self.y + dy) % self.grid_size
        self.x, self.y = new_x, new_y

class Environment:
```

```python
'''
Represents the environment containing agents.
Responsible for updating the agents and visualizing their
↪ states.
'''
def __init__(self, grid_size, num_agents):
    self.grid_size = grid_size
    self.agents = [Agent(i, random.randint(0, grid_size-1),
        ↪ random.randint(0, grid_size-1), grid_size)
                   for i in range(num_agents)]

def update(self):
    '''Update the state of all agents.'''
    for agent in self.agents:
        agent.move()

def visualize(self):
    '''Visualize the current state of the environment.'''
    plt.figure(figsize=(6, 6))
    plt.xlim(0, self.grid_size)
    plt.ylim(0, self.grid_size)
    x_coords = [agent.x for agent in self.agents]
    y_coords = [agent.y for agent in self.agents]
    plt.scatter(x_coords, y_coords)
    plt.grid(True)
    plt.show()

# Simulation parameters
grid_size = 10
num_agents = 5
num_steps = 10

# Initialize the environment
env = Environment(grid_size, num_agents)

# Run the simulation
for step in range(num_steps):
    print(f"Step {step+1}")
    env.update()
    env.visualize()
```

This code defines the fundamental elements necessary for an agent-based model:

- **Agent** class models an individual agent with methods to move randomly within a grid.

- **Environment** class maintains the collection of agents, updates their states, and provides a function to visualize the agents' positions within the boundary grid.

- The simulation loop initializes the `Environment`, updates agent positions, and visualizes their movements over a prescribed number of steps.

This basic agent-based model can be extended to simulate more complex interactions by adding specific rules or objectives for the agents, inter-agent communication, adaptive behavior, or resource management within the environment.

Chapter 27

Protein Structure Prediction with Homology Modeling

Below is a Python code snippet that demonstrates the core computational elements for predicting protein structures based on sequence homology. This includes the use of alignment tools to find homologous sequences, constructing a multiple sequence alignment, and building a homology model.

```
from Bio import AlignIO, SeqIO
from Bio.Align import MultipleSeqAlignment
from Bio.Seq import Seq
from Bio.SeqRecord import SeqRecord
from Bio.PDB import PDBParser, PPBuilder, MMCIFParser
from Bio.PDB.Model import Model
from Bio.PDB.Chain import Chain
from Bio.PDB.Structure import Structure
from Bio.Align.Applications import ClustalwCommandline
import os

def perform_sequence_alignment(input_sequences, output_alignment):
    '''
    Perform multiple sequence alignment using ClustalW.
    :param input_sequences: Path to input file with unaligned
    ↪ sequences.
    :param output_alignment: Path for output alignment file.
    :return: None
    '''
    clustalw_exe = r"/usr/local/bin/clustalw2"   # Path to ClustalW
    ↪ executable
```

```python
    clustalw_cline = ClustalwCommandline(clustalw_exe,
    ↪   infile=input_sequences)
    clustalw_cline()

def build_protein_model(template_structure_path,
↪   sequence_alignment_path, output_model_path):
    '''
    Build homology model of a protein.
    :param template_structure_path: File path to the structure of
    ↪   the template protein.
    :param sequence_alignment_path: File path to the multiple
    ↪   sequence alignment.
    :param output_model_path: File path where the output model will
    ↪   be saved.
    :return: None
    '''
    # Parsing the template structure
    parser = PDBParser(QUIET=True)
    structure = parser.get_structure('Template',
    ↪   template_structure_path)

    # Retrieving alignment from file
    alignment = AlignIO.read(sequence_alignment_path, 'clustal')

    # Placeholder for homology model construction
    # Typically involves software like MODELLER, omitted in this
    ↪   simple illustrative example

    # For demonstration, creating a new model structure with the
    ↪   same atom and chain arrangement
    model = Model('HomologyModel')
    chain = Chain('A')
    for chain_obj in structure.get_chains():
        for residue in chain_obj:
            chain.add(residue.copy())

    structure.add(chain)

    # Writing the new model to a PDB file
    io = PDBIO()
    io.set_structure(structure)
    io.save(output_model_path)

def extract_amino_acid_sequence(structure_path):
    '''
    Extract amino acid sequence of a protein from its structure
    ↪   file.
    :param structure_path: File path to PDB or MMCIF structure file.
    :return: Amino acid sequence as a string.
    '''
    parser = MMCIFParser(QUIET=True) if
    ↪   structure_path.endswith('.cif') else PDBParser(QUIET=True)
    structure = parser.get_structure('Protein', structure_path)
```

```
    ppb = PPBuilder()

    sequence = ""
    for pp in ppb.build_peptides(structure):
        sequence += str(pp.get_sequence())

    return sequence

# Example usage of the process
template_structure = "template.pdb"
unaligned_sequences = "sequences.fasta"
output_alignment = "aligned.aln"
output_model = "homology_model.pdb"

perform_sequence_alignment(unaligned_sequences, output_alignment)
build_protein_model(template_structure, output_alignment,
 ↪  output_model)

amino_acid_sequence =
 ↪  extract_amino_acid_sequence(template_structure)
print("Extracted Amino Acid Sequence:", amino_acid_sequence)
```

This code snippet demonstrates some of the essential steps involved in predicting protein structures based on sequence homology:

- `perform_sequence_alignment` function performs a multiple sequence alignment using ClustalW, preparing sequences for modeling.

- `build_protein_model` simulates the construction of a protein homology model using a template structure and alignment data.

- `extract_amino_acid_sequence` retrieves the amino acid sequence from a given protein structure file.

By leveraging sequence alignment and structural templates, the method allows for the hypothetical creation of protein models that aid in understanding structural biology. Note that the snippet involves placeholders where more sophisticated modeling software would typically be required.

Chapter 28

Molecular Dynamics Simulation Algorithms

Below is a Python code snippet that illustrates the core computational elements of molecular dynamics simulations, focusing on initializing particles, computing forces based on the Lennard-Jones potential, updating particle positions using integration methods, and calculating the system's energy.

```
import numpy as np

class Particle:
    def __init__(self, position, velocity, mass):
        self.position = np.array(position)
        self.velocity = np.array(velocity)
        self.mass = mass

def compute_lennard_jones_force(r, epsilon=1.0, sigma=1.0):
    '''
    Calculate the Lennard-Jones force given the distance.
    :param r: Distance between particles.
    :param epsilon: Depth of the potential well.
    :param sigma: Finite distance at which the inter-particle
    ↪  potential is zero.
    :return: Lennard-Jones force.
    '''
    if r == 0:
        return 0
    # Lennard-Jones potential formula
    force = 4 * epsilon * ((12 * (sigma**12) / r**13) - (6 *
    ↪  (sigma**6) / r**7))
    return force
```

```python
def update_positions(particles, dt):
    '''
    Update positions of particles using basic Verlet integration.
    :param particles: List of Particle objects.
    :param dt: Time step.
    '''
    for particle in particles:
        particle.position += particle.velocity * dt

def update_velocities(particles, forces, dt):
    '''
    Update velocities of particles using basic Verlet integration.
    :param particles: List of Particle objects.
    :param forces: List of forces acting on each particle.
    :param dt: Time step.
    '''
    for i, particle in enumerate(particles):
        particle.velocity += forces[i] / particle.mass * dt

def compute_forces(particles):
    '''
    Compute the forces acting on each particle.
    :param particles: List of Particle objects.
    :return: List of forces for each particle.
    '''
    num_particles = len(particles)
    forces = [np.zeros(3) for _ in range(num_particles)]
    for i in range(num_particles):
        for j in range(i+1, num_particles):
            r_vec = particles[j].position - particles[i].position
            r = np.linalg.norm(r_vec)
            force_magnitude = compute_lennard_jones_force(r)
            force_vector = force_magnitude * r_vec / r
            forces[i] += force_vector
            forces[j] -= force_vector
    return forces

def compute_kinetic_energy(particles):
    '''
    Calculate the total kinetic energy of the system.
    :param particles: List of Particle objects.
    :return: Total kinetic energy.
    '''
    kinetic_energy = sum(0.5 * p.mass * np.dot(p.velocity,
        ↪ p.velocity) for p in particles)
    return kinetic_energy

def simulate_molecular_dynamics(particles, dt, num_steps):
    '''
    Run the molecular dynamics simulation.
    :param particles: List of Particle objects.
    :param dt: Time step.
```

```
    :param num_steps: Number of simulation steps.
    '''
    for step in range(num_steps):
        forces = compute_forces(particles)
        update_positions(particles, dt)
        update_velocities(particles, forces, dt)

# Example usage
particles = [
    Particle(position=[0, 0, 0], velocity=[0.1, 0.2, 0.3],
        mass=1.0),
    Particle(position=[1, 0, 0], velocity=[-0.1, 0.0, 0.1],
        mass=1.0)
]
dt, num_steps = 0.01, 1000

simulate_molecular_dynamics(particles, dt, num_steps)
kinetic_energy = compute_kinetic_energy(particles)
print("Total Kinetic Energy:", kinetic_energy)
```

This code defines several key functions necessary for performing molecular dynamics simulations with basic Lennard-Jones force calculations:

- `compute_lennard_jones_force` calculates the force between two particles using the Lennard-Jones potential.

- `update_positions` and `update_velocities` update the particles' positions and velocities using Verlet integration for numerical stability.

- `compute_forces` computes pairwise forces between particles to simulate interactions.

- `compute_kinetic_energy` calculates the total kinetic energy of the system, useful for monitoring simulation stability.

- `simulate_molecular_dynamics` is the main simulation loop executing the defined steps over several iterations.

The final block demonstrates initializing a simple two-particle system and performing the simulation, printing the total kinetic energy at completion.

Chapter 29

Protein Docking Algorithms

Below is a Python code snippet that encompasses the core computational elements of protein-protein docking, including the calculation of binding affinities and scoring functions, pose generation, and evaluation of interaction energies.

```python
import numpy as np
from scipy.optimize import minimize

def calculate_affinity(protein, ligand):
    '''
    Calculate the affinity score between a protein and a ligand.
    :param protein: 3D coordinates of the protein's binding site.
    :param ligand: 3D coordinates of the ligand.
    :return: Affinity score.
    '''
    distance = np.linalg.norm(protein - ligand)
    return np.exp(-distance)  # Simplified affinity model

def docking_score(protein, ligand, conformation):
    '''
    Compute the docking score for a given protein-ligand
        conformation.
    :param protein: 3D coordinates of the protein's binding site.
    :param ligand: 3D coordinates of the ligand.
    :param conformation: Rotation and translation parameters.
    :return: Docking score.
    '''
    # Apply rotation and translation to ligand
    transformed_ligand = apply_transformation(ligand, conformation)
    affinity = calculate_affinity(protein, transformed_ligand)
```

```python
        penalty = np.sum(np.square(conformation[:3]))  # Penalize large
        ↪ transformations
        return -affinity + penalty

    def apply_transformation(ligand, conformation):
        '''
        Apply rotation and translation to a ligand.
        :param ligand: 3D coordinates of the ligand.
        :param conformation: The transformation parameters.
        :return: Transformed 3D coordinates.
        '''
        rotation_matrix = np.eye(3)  # Simplified as identity for
        ↪ demonstration
        translation_vector = conformation[3:]
        return np.dot(ligand, rotation_matrix) + translation_vector

    def optimize_docking(protein, ligand):
        '''
        Optimize protein-ligand docking by minimizing the docking score.
        :param protein: 3D coordinates of the protein's binding site.
        :param ligand: 3D coordinates of the ligand.
        :return: Optimal conformation parameters.
        '''
        initial_conformation = np.zeros(6)  # Initial guess for rotation
        ↪ and translation
        result = minimize(docking_score, initial_conformation,
        ↪ args=(protein, ligand))
        return result.x

    # Example protein and ligand coordinates
    protein_coords = np.array([1.0, 2.0, 3.0])
    ligand_coords = np.array([2.5, 3.0, 4.0])

    # Optimize docking
    optimal_conformation = optimize_docking(protein_coords,
    ↪ ligand_coords)
    transformed_ligand = apply_transformation(ligand_coords,
    ↪ optimal_conformation)
    final_score = docking_score(protein_coords, ligand_coords,
    ↪ optimal_conformation)

    print("Optimal Conformation:", optimal_conformation)
    print("Final Docking Score:", final_score)
```

This code defines several key functions necessary for protein-protein docking:

- `calculate_affinity` function computes a simplified affinity score based on the distance between the protein and ligand.

- `docking_score` calculates the overall docking score, combining affinity with a penalty for large conformational transformations.

- `apply_transformation` applies rotation and translation to the ligand coordinates based on the conformation parameters.

- `optimize_docking` uses optimization to find the optimum conformation, minimizing the docking score.

The final block of code provides a simple example of optimizing docking between a protein and ligand using dummy 3D coordinates.

Chapter 30

Cryo-EM Data Processing Algorithms

Below is a Python code snippet that encompasses the core computational elements for processing cryo-electron microscopy data, including image preprocessing, 3D reconstruction using iterative alignment, and structure refinement.

```python
import numpy as np
import scipy.ndimage as ndi
from skimage.restoration import denoise_wavelet
from sklearn.decomposition import IncrementalPCA

def preprocess_image(image):
    '''
    Preprocess cryo-EM image data with noise reduction and
       normalization.
    :param image: Input raw image data.
    :return: Preprocessed image.
    '''
    # Denoise the image using wavelet transform
    denoised_image = denoise_wavelet(image, method='BayesShrink',
       mode='soft')
    # Normalize the image
    normalized_image = (denoised_image - np.mean(denoised_image)) /
       np.std(denoised_image)
    return normalized_image

def iterative_alignment(images, initial_guess, max_iterations=10):
    '''
    Perform iterative alignment and averaging of images to improve
       3D map.
    :param images: A list of 2D cryo-EM images.
```

```python
    :param initial_guess: Initial 3D model guess.
    :param max_iterations: Number of iterations for alignment.
    :return: Refined 3D map.
    '''
    current_map = initial_guess
    for iteration in range(max_iterations):
        aligned_sum = np.zeros_like(current_map)
        for image in images:
            # Compute alignment parameters
            alignment_parameters = compute_alignment(image,
            ↪    current_map)
            # Align image to current map
            aligned_image = align_image(image, alignment_parameters)
            # Sum aligned images to update 3D map
            aligned_sum += aligned_image
        current_map = aligned_sum / len(images)  # Average aligned
        ↪    images
    return current_map

def compute_alignment(image, map3d):
    '''
    Placeholder function to compute alignment parameters.
    :param image: 2D cryo-EM image.
    :param map3d: Current 3D model map.
    :return: Alignment parameters.
    '''
    # Dummy alignment parameters calculation
    return np.array([0, 0, 0])  # Example: [x translation, y
    ↪    translation, rotation]

def align_image(image, parameters):
    '''
    Dummy function to align the image based on parameters.
    :param image: 2D cryo-EM image.
    :param parameters: Computed alignment parameters.
    :return: Aligned image.
    '''
    # Apply translation and rotation as per parameters (dummy
    ↪    operation)
    return ndi.shift(ndi.rotate(image, parameters[2],
    ↪    reshape=False), parameters[:2])

def refine_structure(map3d):
    '''
    Perform final refinement on the 3D map.
    :param map3d: The refined 3D map from iterative alignment.
    :return: The final high-resolution structure.
    '''
    # Perform Principal Component Analysis for dimensionality
    ↪    reduction
    ipca = IncrementalPCA(n_components=map3d.shape[0])
    map_reduced = ipca.fit_transform(map3d)
```

```
    # Reconstruct high-resolution map (simple placeholder for
    ↪  demonstration)
    map_final = ipca.inverse_transform(map_reduced)
    return map_final

# Example usage
images = [np.random.rand(256, 256) for _ in range(100)]  # List of
↪  dummy images
initial_guess = np.zeros((30, 256, 256))  # Dummy initial 3D model

# Preprocess images
preprocessed_images = [preprocess_image(img) for img in images]
# Perform iterative alignment
refined_map = iterative_alignment(preprocessed_images,
↪  initial_guess)
# Refine the structure
final_structure = refine_structure(refined_map)

print("Final high-resolution structure shape:",
↪  final_structure.shape)
```

This code defines several key functions necessary for the reconstruction of high-resolution structures from cryo-EM data:

- `preprocess_image` function preprocesses raw image data to enhance quality through noise reduction and normalization.

- `iterative_alignment` aligns and averages multiple cryo-EM images iteratively to refine the 3D reconstruction map.

- `compute_alignment` and `align_image` are placeholders for calculating the alignment parameters and performing the alignment, respectively.

- `refine_structure` uses principal component analysis to further refine the 3D map for achieving high-resolution results.

The final block of code demonstrates the steps of preprocessing, iterative alignment, and structure refinement using dummy data for illustration purposes.

Chapter 31

Optimizing CRISPR Guide RNA Design

Below is a Python code snippet that encompasses the core computational elements for designing guide RNAs with high specificity and efficiency in gene editing, including calculation of on-target scoring, off-target prediction, and final scoring for optimal guide RNA selection.

```python
from Bio.Seq import Seq
import numpy as np

def calculate_on_target_score(guide_rna):
    '''
    Calculate the on-target score for a guide RNA sequence.
    :param guide_rna: Guide RNA sequence as a string.
    :return: On-target score.
    '''
    # Implementing a dummy calculation for illustration
    gc_content = (guide_rna.count('G') + guide_rna.count('C')) / len(guide_rna)
    score = 1.0 if 0.4 <= gc_content <= 0.6 else 0.5
    return score * 100

def predict_off_target_sites(guide_rna, genome_sequence):
    '''
    Predict potential off-target sites for a given guide RNA within
        a genome.
    :param guide_rna: Guide RNA sequence as a string.
    :param genome_sequence: Genome sequence as a string.
    :return: List of potential off-target positions.
    '''
```

```
        # Dummy implementation assumes a simple Hamming distance check
        positions = []
        for i in range(len(genome_sequence) - len(guide_rna) + 1):
            window = genome_sequence[i:i+len(guide_rna)]
            if sum(1 for a, b in zip(guide_rna, window) if a != b) <= 2:
                positions.append(i)
        return positions

def calculate_final_score(guide_rna, genome_sequence):
    '''
    Calculate the final score combining on-target and off-target
    ↪ predictions.
    :param guide_rna: Guide RNA sequence as a string.
    :param genome_sequence: Genome sequence as a string.
    :return: Final score for guide RNA efficacy.
    '''
    on_target_score = calculate_on_target_score(guide_rna)
    off_target_sites = predict_off_target_sites(guide_rna,
    ↪ genome_sequence)

    # Heuristic scoring: penalize for off-target matches
    penalty = len(off_target_sites) * 10
    final_score = on_target_score - penalty
    return max(0, final_score)

# Example use of the functions
guide_rna_seq = "AGCTAGCTAGCTAGCTAGCTA"
genome_seq =
↪ "TTAGCTAGCTAGCTAGCTAGCTAGCTTTAGCTAGCTAGCTAGCTAGCTAGCTAGCTAGCT"

on_target = calculate_on_target_score(guide_rna_seq)
off_targets = predict_off_target_sites(guide_rna_seq, genome_seq)
final = calculate_final_score(guide_rna_seq, genome_seq)

print("On-Target Score:", on_target)
print("Off-Target Sites:", off_targets)
print("Final gRNA Score:", final)
```

This code defines several key functions necessary for evaluating guide RNA efficiency and specificity:

- `calculate_on_target_score` computes the on-target score based on guide RNA GC content.

- `predict_off_target_sites` uses a simple heuristic to predict off-target binding sites in a genome.

- `calculate_final_score` integrates both on-target and off-target scores to determine the final efficacy of the guide RNA.

The final block of code provides examples of computing these elements using dummy data to illustrate the process of guide RNA design.

Chapter 32

Off-Target Prediction in Genome Editing

Below is a Python code snippet that encompasses the core computational elements for predicting potential off-target sites in genome editing applications, including guide RNA design and off-target analysis.

```
import numpy as np
import pandas as pd
from Bio.Seq import Seq
from Bio.SeqUtils import nt_search
from sklearn.metrics.pairwise import cosine_similarity

def find_potential_off_targets(genome, guide_rna,
    mismatch_tolerance=3):
    '''
    Identifies potential off-target sites in the genome.
    :param genome: DNA sequence of the genome.
    :param guide_rna: gRNA sequence.
    :param mismatch_tolerance: Number of allowed mismatches.
    :return: List of potential off-target sites.
    '''
    off_target_sites = []
    guide_length = len(guide_rna)
    for i in range(len(genome) - guide_length + 1):
        window = genome[i:i+guide_length]
        mismatches = sum(1 for a, b in zip(window, guide_rna) if a
            != b)
        if mismatches <= mismatch_tolerance:
            off_target_sites.append((i, window, mismatches))
    return off_target_sites
```

```python
def calculate_similarity_matrix(sequences):
    '''
    Calculate the similarity matrix using cosine similarity.
    :param sequences: List of sequences to compare.
    :return: Similarity matrix.
    '''
    encoded_data = [np.array([ord(char) for char in seq]) for seq in
    ↪  sequences]
    similarity_matrix = cosine_similarity(encoded_data)
    return similarity_matrix

def guide_rna_design(target_sequence, pam_pattern="NGG"):
    '''
    Designs guide RNA based on a target sequence and PAM pattern.
    :param target_sequence: DNA sequence where CRISPR targets.
    :param pam_pattern: PAM pattern for CRISPR.
    :return: Designed guide RNA sequences.
    '''
    search_result = nt_search(str(target_sequence), pam_pattern)
    guide_rnas = [str(target_sequence[max(0, pos - 20):pos]) for pos
    ↪  in search_result[1:]]
    return guide_rnas

# Define genome and guide RNA
genome_sequence =
↪  "AATCGTAGCGTACGATCGTAGCTAGGGCGTACGATCGTACGTACGTACG"
guide_rna_example = "CGTACGTACGTACGTACGTA"

# Find potential off-target sites
off_targets = find_potential_off_targets(genome_sequence,
↪  guide_rna_example)

# Calculate similarity for off-target evaluation
sequences_to_compare = [seq for _, seq, _ in off_targets]
similarity_matrix =
↪  calculate_similarity_matrix(sequences_to_compare)

# Design potential guide RNAs
guide_rnas = guide_rna_design(Seq(genome_sequence))

print("Potential off-target sites and their mismatches:",
↪  off_targets)
print("Similarity matrix of off-target sequences:",
↪  similarity_matrix)
print("Designed guide RNAs:", guide_rnas)
```

This code defines several key functions necessary for predicting off-target effects in genome editing applications:

- `find_potential_off_targets` function identifies possible off-target sites in a provided genome sequence based on a specified guide RNA.

- `calculate_similarity_matrix` computes the similarity matrix of potential off-target sequences by using cosine similarity, providing insights into sequence resemblance.

- `guide_rna_design` proposes guide RNA sequences from a target sequence by locating PAM sites, which are crucial for CRISPR targeting.

The final portion of the code demonstrates the execution of these functions to identify off-target effects and design guide RNAs for a mock genome sequence.

Chapter 33

Homology-Directed Repair Simulation

Below is a Python code snippet that encompasses the core computational elements for simulating the outcomes of homology-directed repair (HDR) during genome editing.

```python
import numpy as np
import random

def simulate_hdr(target_dna, donor_dna, cut_site,
    repair_efficiency=0.5, num_simulations=1000):
    '''
    Simulate Homology-Directed Repair (HDR) outcomes.
    :param target_dna: Original DNA sequence where the cut occurs.
    :param donor_dna: DNA template used for repair.
    :param cut_site: Index where the cut is made in the target DNA.
    :param repair_efficiency: Probability of HDR occurring instead
        of non-homologous end joining (NHEJ).
    :param num_simulations: Number of simulations to run.
    :return: A dictionary with frequencies of possible sequences.
    '''
    outcomes = {
        'HDR': 0,
        'NHEJ': 0,
        'original': 0
    }

    for _ in range(num_simulations):
        if random.random() < repair_efficiency:
            # Simulate HDR event
            repaired_sequence = target_dna[:cut_site] + donor_dna +
                target_dna[cut_site:]
```

```
            outcomes['HDR'] += 1
    else:
        # Simulate NHEJ or other repair event leading to a
        ↪ sequence change
        if random.random() < 0.1:   # Assume 10% of NHEJ leads to
        ↪ original sequence
            outcomes['original'] += 1
        else:
            outcomes['NHEJ'] += 1

# Normalize the counts to frequencies
for key in outcomes.keys():
    outcomes[key] /= num_simulations

return outcomes

# Example usage
target_dna = "ATCGATCGATCG"
donor_dna = "GGGCCC"
cut_site = 6
repair_outcomes = simulate_hdr(target_dna, donor_dna, cut_site)

print("Simulated HDR Outcomes:")
print("HDR Frequency:", repair_outcomes['HDR'])
print("NHEJ Frequency:", repair_outcomes['NHEJ'])
print("Original Frequency:", repair_outcomes['original'])
```

This code defines several key functions and simulations for homology-directed repair (HDR):

- `simulate_hdr` function models the process of HDR by simulating DNA repair outcomes given a cut site and donor template.

- The function uses parameters like repair efficiency and number of simulations to determine the frequency of different repair outcomes.

- The given example illustrates using this function with specific DNA sequences and a cut site to predict the repair outcome frequencies.

This snippet is a basic yet extendable framework for exploring the efficiency and accuracy of HDR in genome editing applications.

Chapter 34

Gene Expression Quantification Algorithms

Below is a Python code snippet that encompasses the core computational elements for quantifying gene expression levels from transcriptomic data, including normalization, differential expression analysis, and visualization.

```
import pandas as pd
import numpy as np
from sklearn.preprocessing import StandardScaler
import matplotlib.pyplot as plt
import seaborn as sns
import statsmodels.api as sm

# Load transcriptomic data
def load_data(file_path):
    '''
    Load gene expression data from a file.
    :param file_path: Path to the data file.
    :return: pandas DataFrame of expression levels.
    '''
    return pd.read_csv(file_path, sep='\t', index_col=0)

# Normalize data using z-score normalization
def z_score_normalization(data):
    '''
    Apply z-score normalization to gene expression data.
    :param data: DataFrame of gene expression levels.
    :return: Normalized DataFrame.
```

```python
    scaler = StandardScaler()
    normalized_data = scaler.fit_transform(data)
    return pd.DataFrame(normalized_data, index=data.index,
    ↪    columns=data.columns)

# Perform differential expression analysis
def differential_expression_analysis(normalized_data, group_labels):
    '''
    Conduct differential expression analysis using linear models.
    :param normalized_data: Z-score normalized DataFrame.
    :param group_labels: Labels for experimental groups.
    :return: DataFrame with differential expression results.
    '''
    results = []
    for gene in normalized_data.index:
        model = sm.OLS(normalized_data.loc[gene],
        ↪    group_labels).fit()
        p_value = model.pvalues[1]
        results.append({'Gene': gene, 'p_value': p_value})
    return pd.DataFrame(results).set_index('Gene')

# Visualization using heatmaps
def visualize_data_heatmap(normalized_data, title='Heatmap of Gene
↪    Expression'):
    '''
    Visualize gene expression data using a heatmap.
    :param normalized_data: DataFrame of normalized data.
    :param title: Title of the plot.
    '''
    plt.figure(figsize=(10, 8))
    sns.heatmap(normalized_data, cmap='viridis', cbar_kws={'label':
    ↪    'Expression'})
    plt.title(title)
    plt.xlabel('Samples')
    plt.ylabel('Genes')
    plt.show()

# Example usage:
# Load data
data = load_data('transcriptomic_data.tsv')

# Normalize data
normalized_data = z_score_normalization(data)

# Group labels for differential expression analysis (example labels)
group_labels = np.array([0, 0, 1, 1, 1])  # 0 for control, 1 for
↪    treated

# Perform differential expression analysis
de_results = differential_expression_analysis(normalized_data,
↪    group_labels)
```

```
# Visualize normalized data
visualize_data_heatmap(normalized_data, title='Heatmap of Normalized
↪    Gene Expression')

print(de_results.head())
```

This code defines several functions necessary for quantifying gene expression levels:

- `load_data` function reads and organizes gene expression data from a file.

- `z_score_normalization` applies z-score normalization to ensure comparability across samples.

- `differential_expression_analysis` performs statistical analysis to identify differentially expressed genes between experimental groups.

- `visualize_data_heatmap` provides a visual representation of gene expression data using a heatmap.

The latter part of the code demonstrates loading, normalizing, analyzing, and visualizing transcriptomic data with example group labels.

Chapter 35

Single-Cell RNA Sequencing Data Analysis

Below is a Python code snippet that encompasses essential algorithms for processing and analyzing gene expression data from single cells, including data normalization, clustering, and visualization.

```
import numpy as np
import pandas as pd
from sklearn.preprocessing import StandardScaler
from sklearn.decomposition import PCA
from sklearn.cluster import KMeans
from matplotlib import pyplot as plt
import seaborn as sns

# Load single-cell RNA sequencing data
def load_data(file_path):
    '''
    Load gene expression data from a file.
    :param file_path: Path to the data file.
    :return: Pandas DataFrame containing the expression data.
    '''
    return pd.read_csv(file_path, sep='\t', index_col=0)

# Normalize the data
def normalize_data(df):
    '''
    Normalize the data using z-score standardization.
    :param df: Pandas DataFrame of expression values.
```

```python
    :return: Normalized data as a Numpy array.
    '''
    scaler = StandardScaler()
    return scaler.fit_transform(df)

# Perform dimensionality reduction using PCA
def perform_pca(data, n_components=2):
    '''
    Perform PCA to reduce data dimensionality.
    :param data: Normalized Numpy array of data.
    :param n_components: Number of principal components to retain.
    :return: PCA-transformed data and explained variance ratio.
    '''
    pca = PCA(n_components=n_components)
    principal_components = pca.fit_transform(data)
    return principal_components, pca.explained_variance_ratio_

# Cluster the data using KMeans
def cluster_data(data, n_clusters):
    '''
    Cluster data using KMeans algorithm.
    :param data: Data to cluster (e.g., PCA-transformed data).
    :param n_clusters: Number of clusters.
    :return: Cluster labels for each data point.
    '''
    kmeans = KMeans(n_clusters=n_clusters, random_state=0)
    return kmeans.fit_predict(data)

# Visualize the clustered data
def plot_clusters(pca_data, cluster_labels, explained_variance):
    '''
    Plot the PCA-transformed data colored by cluster label.
    :param pca_data: PCA-transformed data for plotting.
    :param cluster_labels: Cluster labels obtained from clustering.
    :param explained_variance: Explained variance of the PCA
    ↪ components.
    '''
    plt.figure(figsize=(10, 7))
    sns.scatterplot(x=pca_data[:, 0], y=pca_data[:, 1],
    ↪ hue=cluster_labels, palette='viridis', legend='full')
    plt.title('PCA of Single-Cell Data - Explained Variance: PC1
    ↪ {0:.2f}%, PC2 {1:.2f}%'.format(explained_variance[0] * 100,
    ↪ explained_variance[1] * 100))
    plt.xlabel('Principal Component 1')
    plt.ylabel('Principal Component 2')
    plt.show()

# Main pipeline execution
if __name__ == "__main__":
    # File path to the gene expression data
    file_path = 'single_cell_expression_data.tsv'

    # Load and preprocess the data
```

```
expression_data = load_data(file_path)
normalized_data = normalize_data(expression_data)

# PCA transformation
pca_result, variance_ratio = perform_pca(normalized_data)

# Clustering
cluster_labels = cluster_data(pca_result, n_clusters=3)

# Visualization
plot_clusters(pca_result, cluster_labels, variance_ratio)
```

This code defines several key functions necessary for single-cell RNA-seq data analysis:

- `load_data` function loads the gene expression data from a specified file.

- `normalize_data` performs z-score standardization to normalize gene expression values.

- `perform_pca` performs dimensionality reduction using PCA, retaining principal components.

- `cluster_data` applies KMeans clustering to identify distinct groups within the data.

- `plot_clusters` visualizes the PCA-transformed data, coloring each point by its cluster label.

The main pipeline integrates these steps to process and analyze single-cell RNA sequencing data, enabling meaningful biological interpretation.

Chapter 36

Clustering Techniques for Cellular Heterogeneity

Below is a Python code snippet that applies clustering algorithms to identify subpopulations within heterogeneous cell samples. The code utilizes the popular k-means clustering algorithm to detect clusters in high-dimensional single-cell RNA sequencing data and visualizes the results using t-distributed stochastic neighbor embedding (t-SNE) for dimensionality reduction.

```python
import numpy as np
from sklearn.cluster import KMeans
from sklearn.manifold import TSNE
import matplotlib.pyplot as plt

def preprocess_data(single_cell_data):
    '''
    Preprocess the single-cell data by normalizing the gene
    ↪ expression levels.
    :param single_cell_data: 2D array where each row is a cell, and
    ↪ each column is a gene.
    :return: Normalized data.
    '''
    # Normalize data so that expression levels are in range [0, 1]
    return single_cell_data / single_cell_data.max(axis=0)

def apply_kmeans_clustering(preprocessed_data, num_clusters):
    '''
    Apply k-means clustering to the preprocessed single-cell data.
```

```
    :param preprocessed_data: Normalized single-cell data.
    :param num_clusters: Number of clusters (subpopulations) to
    ↪    identify.
    :return: Array of cluster labels for each cell.
    '''
    kmeans = KMeans(n_clusters=num_clusters, random_state=42)
    return kmeans.fit_predict(preprocessed_data)

def visualize_clusters(preprocessed_data, cluster_labels):
    '''
    Visualize the clustering result using t-SNE.
    :param preprocessed_data: Normalized single-cell data.
    :param cluster_labels: Cluster labels for each cell from
    ↪    k-means.
    '''
    # Reduce dimensions for visualization
    tsne = TSNE(n_components=2, perplexity=30, random_state=42)
    reduced_data = tsne.fit_transform(preprocessed_data)

    # Plotting
    plt.figure(figsize=(10, 8))
    scatter = plt.scatter(reduced_data[:, 0], reduced_data[:, 1],
    ↪    c=cluster_labels, cmap='viridis', alpha=0.7)

    # Add legend
    plt.legend(*scatter.legend_elements(), title="Clusters")
    plt.title('t-SNE visualization of clustered single-cell data')
    plt.xlabel('t-SNE Component 1')
    plt.ylabel('t-SNE Component 2')
    plt.show()

# Example usage with dummy data
np.random.seed(0)
dummy_data = np.random.rand(100, 50)  # 100 cells, 50 genes

# Preprocess the data
normalized_data = preprocess_data(dummy_data)

# Apply clustering
cluster_labels = apply_kmeans_clustering(normalized_data,
↪    num_clusters=3)

# Visualize the results
visualize_clusters(normalized_data, cluster_labels)
```

This code snippet defines essential functions for clustering analysis in single-cell genomics:

- `preprocess_data` normalizes the gene expression data to prepare it for clustering.

- `apply_kmeans_clustering` applies the k-means algorithm to cluster cells into distinct subpopulations.

- `visualize_clusters` uses t-SNE to reduce the dimensionality of the clustered data and matplotlib for visualization, giving insights into the distribution and separation of cell populations.

In the example provided, dummy data represents single-cell expression profiles, which the code processes, clusters, and visualizes effectively.

Chapter 37

Dimensionality Reduction in Genomic Data

Below is a Python code snippet that utilizes the t-SNE and UMAP techniques for visualizing and interpreting high-dimensional genomic data, particularly demonstrating their application on synthetic genomic data.

```
import numpy as np
import matplotlib.pyplot as plt
from sklearn.manifold import TSNE
import umap

# Generate synthetic high-dimensional genomic data
def generate_synthetic_data(samples=100, features=50):
    '''
    Generates synthetic genomic data for demonstration purposes.
    :param samples: Number of samples to generate.
    :param features: Number of features per sample.
    :return: A tuple containing data and labels.
    '''
    np.random.seed(42)
    data = np.random.rand(samples, features)
    labels = np.random.choice(['Type 1', 'Type 2'], size=samples)
    return data, labels

# Visualize using t-SNE
def visualize_tsne(data, labels):
    '''
    Visualizes data using t-SNE.
```

```
    :param data: High-dimensional data.
    :param labels: Corresponding labels for coloring.
    '''
    tsne = TSNE(n_components=2, random_state=42)
    tsne_results = tsne.fit_transform(data)

    plt.figure(figsize=(10, 5))
    plt.scatter(tsne_results[:, 0], tsne_results[:, 1], c=[1 if lbl
    ↪    == 'Type 1' else 0 for lbl in labels], cmap='viridis')
    plt.title('t-SNE Visualization')
    plt.colorbar()
    plt.show()

# Visualize using UMAP
def visualize_umap(data, labels):
    '''
    Visualizes data using UMAP.
    :param data: High-dimensional data.
    :param labels: Corresponding labels for coloring.
    '''
    reducer = umap.UMAP(n_components=2, random_state=42)
    umap_results = reducer.fit_transform(data)

    plt.figure(figsize=(10, 5))
    plt.scatter(umap_results[:, 0], umap_results[:, 1], c=[1 if lbl
    ↪    == 'Type 1' else 0 for lbl in labels], cmap='viridis')
    plt.title('UMAP Visualization')
    plt.colorbar()
    plt.show()

# Main demonstration function
def main():
    data, labels = generate_synthetic_data()
    visualize_tsne(data, labels)
    visualize_umap(data, labels)

if __name__ == "__main__":
    main()
```

This code encompasses the implementation of dimensionality reduction techniques t-SNE and UMAP for visualizing high-dimensional genomic data:

- `generate_synthetic_data` function creates synthetic genomic data for visualization demonstrations.

- `visualize_tsne` uses the t-SNE algorithm to reduce dimensions and visually represent the data.

- `visualize_umap` employs the UMAP algorithm to perform similar dimensionality reduction and visualization.

The `main` function demonstrates how these functions work together, generating synthetic data and creating plots that depict the results of t-SNE and UMAP transformations. These results can be used to identify patterns and clusters within the genomic data.

Chapter 38

Gene Regulatory Network Inference

Below is a Python code snippet that demonstrates the process of constructing gene regulatory networks from expression data. This code entails key computational components such as data preprocessing, network inference using correlation metrics, and visualization of the resulting regulatory network.

```python
import pandas as pd
import numpy as np
import networkx as nx
import matplotlib.pyplot as plt
from sklearn.preprocessing import StandardScaler
from scipy.stats import pearsonr

def load_expression_data(file_path):
    """
    Load gene expression data from a CSV file.
    :param file_path: Path to CSV file containing gene expression
        data.
    :return: DataFrame of gene expression data.
    """
    return pd.read_csv(file_path)

def preprocess_expression_data(expression_data):
    """
    Standardize the gene expression data.
    :param expression_data: DataFrame of gene expression data.
    :return: Standardized DataFrame of gene expression data.
    """
    scaler = StandardScaler()
```

```python
        scaled_data = scaler.fit_transform(expression_data)
        return pd.DataFrame(scaled_data,
            ↪ columns=expression_data.columns)

    def infer_network(expression_data, threshold=0.7):
        """
        Infer a gene regulatory network from expression data using
        ↪ Pearson correlation.
        :param expression_data: DataFrame of standardized gene
        ↪ expression data.
        :param threshold: Correlation threshold for inferring edges.
        :return: NetworkX graph object representing the gene regulatory
        ↪ network.
        """
        num_genes = expression_data.shape[1]
        network = nx.Graph()

        for i in range(num_genes):
            for j in range(i + 1, num_genes):
                gene1, gene2 = expression_data.columns[i],
                ↪ expression_data.columns[j]
                corr, _ = pearsonr(expression_data[gene1],
                ↪ expression_data[gene2])

                if abs(corr) >= threshold:
                    network.add_edge(gene1, gene2, weight=corr)

        return network

    def visualize_network(network):
        """
        Visualize the gene regulatory network.
        :param network: NetworkX graph object representing the gene
        ↪ regulatory network.
        """
        pos = nx.spring_layout(network)
        plt.figure(figsize=(12, 12))
        nx.draw(network, pos, with_labels=True, node_size=500,
        ↪ edge_color='b', node_color='c', font_size=10)
        plt.title('Gene Regulatory Network')
        plt.show()

    # Example usage
    file_path = "gene_expression_data.csv"   # Path to your expression
    ↪ data file
    expression_data = load_expression_data(file_path)
    processed_data = preprocess_expression_data(expression_data)
    regulatory_network = infer_network(processed_data)
    visualize_network(regulatory_network)
```

This code provides an example of constructing a gene regulatory network from gene expression data:

- `load_expression_data` function reads the gene expression data from a CSV file.

- `preprocess_expression_data` standardizes the data, facilitating unbiased correlation computation.

- `infer_network` utilizes the Pearson correlation technique to infer regulatory connections, forming network edges.

- `visualize_network` plots the constructed network, highlighting the inferred regulatory interactions among genes.

This workflow shows how regulatory networks can be inferred to determine potential interactions between genes based on their expression levels.

Chapter 39

Transcription Factor Binding Site Prediction

Below is a Python code snippet that encompasses the core computational elements of transcription factor binding site prediction, including data processing, motif scanning, and scoring algorithms.

```
import numpy as np
import pandas as pd

def read_pwm(file_path):
    '''
    Reads a position weight matrix (PWM) from a file.
    :param file_path: Path to the PWM file.
    :return: A dictionary of PWM.
    '''
    pwm = pd.read_csv(file_path, sep='\t', index_col=0)
    return pwm

def score_sequence(sequence, pwm):
    '''
    Computes the binding score for a sequence using a PWM.
    :param sequence: DNA sequence to score.
    :param pwm: Position weight matrix.
    :return: Total binding score for the sequence.
    '''
    score = 0.0
    for i, nucleotide in enumerate(sequence):
        score += pwm.at[nucleotide, i]
    return score

def find_best_match(dna_sequence, pwm):
    '''
```

```
    Finds the best matching subsequence in a DNA sequence according
 ↪  to the PWM.
    :param dna_sequence: The full DNA sequence.
    :param pwm: Position weight matrix.
    :return: The best matching subsequence and its score.
    '''
    k = pwm.shape[1]
    best_score = float('-inf')
    best_subseq = None
    for i in range(len(dna_sequence) - k + 1):
        subseq = dna_sequence[i:i+k]
        current_score = score_sequence(subseq, pwm)
        if current_score > best_score:
            best_score = current_score
            best_subseq = subseq
    return best_subseq, best_score

# Example PWM data and DNA sequence
dna_sequence = "ACGTGGGTGACGTAGTACGCCGA"
pwm_data = {
    'A': [0.1, 0.2, 0.3, 0.4],
    'C': [0.2, 0.3, 0.2, 0.1],
    'G': [0.3, 0.2, 0.2, 0.4],
    'T': [0.4, 0.3, 0.3, 0.1],
}
pwm = pd.DataFrame(pwm_data)

# Running the binding site prediction
best_subsequence, best_score = find_best_match(dna_sequence, pwm)

print("Best matching subsequence:", best_subsequence)
print("Best match score:", best_score)
```

This code defines several key functions for evaluating transcription factor binding sites within a DNA sequence:

- `read_pwm` function reads a position weight matrix from a file that represents the binding preferences of a transcription factor.

- `score_sequence` computes the binding score of a given DNA sequence using the position weight matrix.

- `find_best_match` scans the entire DNA sequence to find and score subsequences, identifying the most likely binding site.

The example given demonstrates how to score a sequence and identify the best binding subsequence using a given PWM, which aids in locating transcription factor binding sites in genetic data.

Chapter 40

Chromatin Accessibility Analysis

Below is a Python code snippet that encompasses the core computational elements for analyzing chromatin accessibility and its impact on gene regulation using machine learning techniques, specifically focusing on the preprocessing of the data, feature extraction, model training, and evaluation.

```python
import numpy as np
import pandas as pd
from sklearn.model_selection import train_test_split
from sklearn.ensemble import RandomForestClassifier
from sklearn.metrics import accuracy_score, roc_auc_score
from scipy.signal import find_peaks

def preprocess_chromatin_data(raw_data):
    '''
    Preprocess chromatin accessibility data.
    :param raw_data: Raw input data containing chromatin
        accessibility measures.
    :return: Preprocessed data ready for analysis.
    '''
    # Normalize the data
    normalized_data = (raw_data - np.min(raw_data)) / \
        (np.max(raw_data) - np.min(raw_data))

    # Feature scaling, e.g., z-score normalization
    mean = np.mean(normalized_data)
    std_dev = np.std(normalized_data)
    standardized_data = (normalized_data - mean) / std_dev
```

```python
    return standardized_data

def extract_features(standardized_data):
    '''
    Extract features such as peak counts and intensities from
    ↪  chromatin data.
    :param standardized_data: Data after preprocessing.
    :return: Feature matrix for model training.
    '''
    # Detect peaks in the chromatin accessibility signal
    peaks, _ = find_peaks(standardized_data, height=0.5)
    peak_count = len(peaks)
    peak_intensities = standardized_data[peaks]

    # Compute summary statistics
    avg_intensity = np.mean(peak_intensities)
    max_intensity = np.max(peak_intensities)

    # Construct feature vector
    features = np.array([peak_count, avg_intensity, max_intensity])

    return features

def train_model(features, labels):
    '''
    Train a Random Forest model to classify chromatin accessibility
    ↪  impact.
    :param features: Feature matrix for training.
    :param labels: Labels for classification.
    :return: Trained model.
    '''
    X_train, X_test, y_train, y_test = train_test_split(features,
    ↪  labels, test_size=0.3, random_state=42)
    model = RandomForestClassifier(n_estimators=100,
    ↪  random_state=42)
    model.fit(X_train, y_train)

    # Evaluate model
    predictions = model.predict(X_test)
    accuracy = accuracy_score(y_test, predictions)
    roc_auc = roc_auc_score(y_test, predictions)

    print(f'Accuracy: {accuracy}, ROC AUC: {roc_auc}')

    return model

# Example dataset
raw_data = np.random.rand(1000)  # Simulated chromatin accessibility
↪  data
labels = np.random.randint(0, 2, 100)  # Simulated binary labels for
↪  chromatin impact

# Preprocess data
```

```
preprocessed_data = preprocess_chromatin_data(raw_data)

# Extract features
features = extract_features(preprocessed_data)

# Reshape features for model training
features_matrix = np.tile(features, (len(labels), 1))  #
      Broadcasting example

# Train model
trained_model = train_model(features_matrix, labels)
```

This code defines several key functions necessary for the analysis of chromatin accessibility and its impact on gene regulation:

- `preprocess_chromatin_data` function normalizes and standardizes chromatin accessibility data.

- `extract_features` identifies peaks and computes summary statistics from chromatin signal data, constructing feature vectors suitable for model input.

- `train_model` trains a Random Forest classifier on the feature matrix to predict the potential impact on gene regulation and evaluates model performance using accuracy and ROC AUC metrics.

The final block of code provides an example of preprocessing simulated chromatin data, extracting features, and training a machine learning model to analyze the data's impact on gene regulation.

Chapter 41

Epigenetic Modification Modeling

Below is a Python code snippet that demonstrates how to model DNA methylation and histone modification patterns using computational algorithms. This example uses simulated data to showcase how such patterns can be analyzed and visualized.

```
import numpy as np
import matplotlib.pyplot as plt
import seaborn as sns

# Simulated data for methylation and histone modifications
np.random.seed(42)
num_sites = 1000
methylation_levels = np.random.beta(2, 5, num_sites)  # Simulated
    beta distribution
histone_modifications = np.random.normal(0.5, 0.1, num_sites)  #
    Normal distribution

def calculate_correlation(methylation_levels,
    histone_modifications):
    '''
    Calculate the correlation between DNA methylation levels and
        histone modifications.
    :param methylation_levels: Array of methylation levels.
    :param histone_modifications: Array of histone modification
        levels.
    :return: Correlation coefficient.
    '''
    return np.corrcoef(methylation_levels, histone_modifications)[0,
        1]
```

```python
def plot_distributions(methylation_levels, histone_modifications):
    '''
    Plot distributions of methylation levels and histone
    ↪ modifications.
    :param methylation_levels: Array of methylation levels.
    :param histone_modifications: Array of histone modification
    ↪ levels.
    '''
    plt.figure(figsize=(12, 6))
    sns.histplot(methylation_levels, color='b', label='Methylation
    ↪ Levels', kde=True)
    sns.histplot(histone_modifications, color='r', label='Histone
    ↪ Modifications', kde=True)
    plt.title("Distributions of DNA Methylation and Histone
    ↪ Modifications")
    plt.legend()
    plt.show()

def analyze_patterns(methylation_levels, histone_modifications):
    '''
    Analyze methylation and histone modification patterns and
    ↪ visualize the correlation.
    :param methylation_levels: Array of methylation levels.
    :param histone_modifications: Array of histone modification
    ↪ levels.
    '''
    plot_distributions(methylation_levels, histone_modifications)
    correlation = calculate_correlation(methylation_levels,
    ↪ histone_modifications)
    plt.figure(figsize=(8, 6))
    sns.scatterplot(x=methylation_levels, y=histone_modifications)
    plt.title(f"Correlation: {correlation:.2f}")
    plt.xlabel("Methylation Levels")
    plt.ylabel("Histone Modifications")
    plt.show()
    return correlation

# Execute analysis
correlation = analyze_patterns(methylation_levels,
↪ histone_modifications)
print(f"Calculated Correlation: {correlation:.2f}")
```

This code incorporates several important functions to model and analyze epigenetic modifications:

- `calculate_correlation` function computes the Pearson correlation coefficient between methylation levels and histone modifications.

- `plot_distributions` visualizes the distributions of methylation levels and histone modifications using histograms and kernel density estimates.

- **analyze_patterns** executes the pattern analysis, generating scatter plots to demonstrate the correlation and visualize the inter-relationship between the two epigenomic modifications.

By running this analysis, researchers can gain insights into the complex interactions between DNA methylation and histone modifications, which are critical for understanding gene regulation processes.

Chapter 42

Non-Coding RNA Function Prediction

The following Python code snippet demonstrates a computational approach to predict the functions of non-coding RNAs, leveraging various computational techniques such as sequence features extraction, machine learning model implementation, and functional prediction.

```python
import numpy as np
from sklearn.feature_extraction.text import CountVectorizer
from sklearn.ensemble import RandomForestClassifier
from sklearn.model_selection import train_test_split
from sklearn.metrics import accuracy_score

def extract_features(sequences):
    '''
    Extract numerical features from RNA sequences using k-mer
    ↪ frequency.
    :param sequences: List of RNA sequences.
    :return: Feature matrix with k-mer frequencies.
    '''
    vectorizer = CountVectorizer(analyzer='char', ngram_range=(3,
    ↪ 3))
    k_mer_matrix = vectorizer.fit_transform(sequences)
    return k_mer_matrix

def load_data():
    '''
    Load and prepare the dataset of sequences and corresponding
    ↪ targets.
    :return: Tuple of sequences and targets.
```

```python
    '''
    # Dummy sequences and labels for demonstration
    sequences = ["AUGCUU", "AUGCGA", "CGUACG", "ACGUAU", "CGCGAA"]
    targets = [0, 1, 0, 1, 0]  # Binary labels representing function
    ↪    classes
    return sequences, targets

def train_model(X, y):
    '''
    Train a Random Forest classifier to predict RNA function.
    :param X: Feature matrix.
    :param y: Target labels.
    :return: Trained model.
    '''
    X_train, X_test, y_train, y_test = train_test_split(X, y,
    ↪    test_size=0.2, random_state=42)
    classifier = RandomForestClassifier(n_estimators=100,
    ↪    random_state=42)
    classifier.fit(X_train, y_train)
    predictions = classifier.predict(X_test)
    accuracy = accuracy_score(y_test, predictions)
    print(f"Model Accuracy: {accuracy:.2f}")
    return classifier

def predict_function(sequence, model, vectorizer):
    '''
    Predict the function class of a new RNA sequence.
    :param sequence: RNA sequence.
    :param model: Trained classification model.
    :param vectorizer: Fitted vectorizer for feature extraction.
    :return: Predicted function class.
    '''
    k_mer_vector = vectorizer.transform([sequence])
    prediction = model.predict(k_mer_vector)
    return prediction[0]

# Main process
sequences, targets = load_data()
features = extract_features(sequences)
vectorizer = CountVectorizer(analyzer='char', ngram_range=(3,
↪    3)).fit(sequences)
model = train_model(features, targets)

# Prediction for demonstration
new_sequence = "CGUAAG"
predicted_function = predict_function(new_sequence, model,
↪    vectorizer)
print(f"Predicted Function for {new_sequence}:
↪    {predicted_function}")
```

This snippet encompasses key functions necessary for functional prediction of non-coding RNAs:

- `extract_features` uses k-mer frequency analysis to transform RNA sequences into a numerical feature matrix.

- `load_data` simulates loading a set of RNA sequences and their associated functional labels.

- `train_model` implements a Random Forest classifier to recognize patterns in sequences corresponding to functional classes.

- `predict_function` applies the trained model and vectorizer to predict the function of a new RNA sequence.

The script concludes with a demonstration of predicting the function of an unseen sequence based on learned features.

Chapter 43

miRNA Target Prediction Algorithms

Below is a Python code snippet that illustrates miRNA target prediction using machine learning techniques, including sequence feature extraction, model training, and prediction of miRNA and mRNA interactions.

```
import numpy as np
import pandas as pd
from sklearn.model_selection import train_test_split
from sklearn.ensemble import RandomForestClassifier
from sklearn.metrics import accuracy_score, classification_report
from Bio.SeqUtils import GC
from Bio.Seq import Seq

def extract_features(mirna_seq, target_seq):
    '''
    Extract features from miRNA and target gene sequences.
    :param mirna_seq: miRNA nucleotide sequence.
    :param target_seq: Target mRNA nucleotide sequence.
    :return: Feature vector.
    '''
    # Example features: length, GC content, and sequence interaction
    features = {
        'mirna_length': len(mirna_seq),
        'target_length': len(target_seq),
        'mirna_gc_content': GC(mirna_seq),
        'target_gc_content': GC(target_seq),
        'sequence_complementarity': complementarity_score(mirna_seq,
          ↪  target_seq),
    }
    return np.array(list(features.values()))
```

```python
def complementarity_score(mirna_seq, target_seq):
    '''
    Computes a simple complementarity-based score for binding
    ↪ potential.
    :param mirna_seq: miRNA nucleotide sequence.
    :param target_seq: Target mRNA nucleotide sequence.
    :return: Complementarity score.
    '''
    score = 0
    min_length = min(len(mirna_seq), len(target_seq))
    for i in range(min_length):
        if mirna_seq[i] == target_seq[i]:
            score += 1
    return score / min_length

# Load example data (replace with real dataset)
data = pd.DataFrame({
    'mirna_seq': ['AUGCUGA', 'UACGUAU', 'CAGUCAU'],
    'target_seq': ['UACGACU', 'GCAUCGU', 'AUGCUGA'],
    'interaction': [1, 0, 1]
})

# Feature extraction
data['features'] = data.apply(lambda row:
↪ extract_features(row['mirna_seq'], row['target_seq']), axis=1)

# Split data into train and test sets
X = np.array(data['features'].tolist())
y = data['interaction']
X_train, X_test, y_train, y_test = train_test_split(X, y,
↪ test_size=0.2, random_state=42)

# Model training
model = RandomForestClassifier(n_estimators=100, random_state=42)
model.fit(X_train, y_train)

# Prediction and evaluation
y_pred = model.predict(X_test)
accuracy = accuracy_score(y_test, y_pred)

print("Accuracy of miRNA target prediction:", accuracy)
print(classification_report(y_test, y_pred))
```

This code snippet performs miRNA target prediction by extracting sequence-based features and leveraging a machine learning model:

- `extract_features` function generates important attributes from miRNA and target sequences, including GC content and a basic complementarity score.

- `complementarity_score` computes a simplification score indicating the potential of miRNA binding to its target.

- The dataset is loaded into a pandas DataFrame and features are extracted for each miRNA-target pair.

- Data is split into training and testing sets to enable model training and evaluation.

- A RandomForestClassifier is used to train the model on extracted features.

- `accuracy_score` and `classification_report` are employed to assess prediction accuracy and provide detailed performance metrics.

Chapter 44

Long Non-Coding RNA Interaction Networks

Below is a Python code snippet that encompasses the core computational elements for constructing interaction networks involving long non-coding RNAs (lncRNAs), including data preprocessing, network construction, and visualization.

```
import numpy as np
import pandas as pd
import networkx as nx
import matplotlib.pyplot as plt
from sklearn.preprocessing import StandardScaler
from scipy.spatial.distance import pdist, squareform

def preprocess_lncrna_data(data_filepath):
    '''
    Load and preprocess lncRNA expression data.
    :param data_filepath: Path to the lncRNA expression dataset.
    :return: Scaled data as a pandas DataFrame.
    '''
    data = pd.read_csv(data_filepath)
    # Assume the data has a format with lncRNAs as columns and
    ↪  samples as rows
    scaler = StandardScaler()
    scaled_data = scaler.fit_transform(data)
    return pd.DataFrame(scaled_data, columns=data.columns,
    ↪   index=data.index)

def construct_interaction_network(lncrna_data, threshold):
```

```
'''
Construct an interaction network of lncRNAs based on expression
↪ correlations.
:param lncrna_data: Preprocessed lncRNA expression data.
:param threshold: Correlation threshold to define edges.
:return: A networkx graph representing the lncRNA interaction
↪ network.
'''
# Compute pairwise correlations
corr_matrix = lncrna_data.corr()
# Apply threshold to define significant interactions
adj_matrix = (corr_matrix >= threshold).astype(int)
np.fill_diagonal(adj_matrix.values, 0)  # Remove self-loops

# Create graph
graph = nx.from_numpy_matrix(adj_matrix.values)
graph = nx.relabel_nodes(graph,
↪    dict(enumerate(lncrna_data.columns)))

return graph

def visualize_interaction_network(graph, title):
    '''
    Visualize the lncRNA interaction network using matplotlib.
    :param graph: The networkx graph object to visualize.
    :param title: Title of the graph.
    '''
    plt.figure(figsize=(12, 12))
    pos = nx.spring_layout(graph)
    nx.draw(graph, pos, with_labels=True, node_size=500,
    ↪    node_color="lightblue", font_size=10, edge_color="gray")
    plt.title(title)
    plt.show()

# Example usage
data_filepath = "lncrna_expression_data.csv"  # Place appropriate
↪ file path
lncrna_data = preprocess_lncrna_data(data_filepath)
lncrna_network = construct_interaction_network(lncrna_data,
↪    threshold=0.7)
visualize_interaction_network(lncrna_network, title="lncRNA
↪    Interaction Network")
```

This code defines several key functions necessary for the implementation and analysis of long non-coding RNA interaction networks:

- `preprocess_lncrna_data` function loads and preprocesses the lncRNA expression data, ensuring that the data is scaled appropriately for correlation analysis.

- `construct_interaction_network` constructs the interaction network based on a specified correlation threshold, using NetworkX for graph representation.

- `visualize_interaction_network` employs matplotlib to visualize the constructed network, providing clear insights into lncRNA interactions.

The final block of code provides an example of how these functions can be employed using a specified dataset, demonstrating the construction and visualization of an lncRNA network.

Chapter 45

Sequence Motif Discovery Algorithms

Below is a Python code snippet that encompasses the core functionalities required for discovering conserved motifs within biological sequences. The algorithms help in identifying common patterns that are biologically significant.

```python
import itertools
import collections

def find_motifs_brute_force(sequences, motif_length):
    '''
    Brute force approach to find all motifs of a given length that are
    common across all sequences.
    :param sequences: List of input DNA/RNA sequences.
    :param motif_length: Length of the motif to find.
    :return: Set of conserved motifs.
    '''
    possible_motifs = set()
    first_seq = sequences[0]

    # Generate all possible motifs from the first sequence
    for i in range(len(first_seq) - motif_length + 1):
        motif = first_seq[i:i + motif_length]
        found_in_all = True

        # Check if this motif is present in every other sequence
        for seq in sequences[1:]:
            if motif not in seq:
                found_in_all = False
                break
```

```python
        if found_in_all:
            possible_motifs.add(motif)

    return possible_motifs

def build_k_mer_histogram(sequences, k):
    '''
    Builds a histogram (frequency count) of all k-mers in the
    ↪ sequences.
    :param sequences: List of input DNA/RNA sequences.
    :param k: Length of the k-mer.
    :return: Histogram dictionary of k-mer counts.
    '''
    k_mer_count = collections.defaultdict(int)

    for seq in sequences:
        for i in range(len(seq) - k + 1):
            k_mer_count[seq[i:i + k]] += 1

    return k_mer_count

# Example sequences to search
sequences = [
    "ATCGAATCG",
    "GGAATCGTA",
    "CCGAATCGG"
]

# Define motif length to search for
motif_length = 3

# Find motifs using brute force method
found_motifs = find_motifs_brute_force(sequences, motif_length)
print("Found Motifs:", found_motifs)

# Build a histogram of k-mers and print it
k_mer_histogram = build_k_mer_histogram(sequences, motif_length)
print("K-mer Histogram:", k_mer_histogram)
```

This code defines several key functions necessary for motif discovery:

- **find_motifs_brute_force**: Searches for all motifs of a given length that appear in all sequences using a brute-force approach.

- **build_k_mer_histogram**: Constructs a histogram of k-mer occurrences across all sequences, which helps identify frequent patterns.

By using these functions, one can systematically identify motifs and generate frequency counts of k-mers within a given set of biological sequences.

Chapter 46

Genome Annotation Pipelines

Below is a Python code snippet that implements a basic workflow for annotating genomic features, using the `Biopython` library to parse genomic sequence data and generate annotations.

```
from Bio import SeqIO
from Bio.SeqFeature import SeqFeature, FeatureLocation

# Function to parse a FASTA file and extract sequences
def parse_fasta(file_path):
    '''
    Parse a FASTA file.
    :param file_path: Path to the FASTA file.
    :return: List of sequence records.
    '''
    return list(SeqIO.parse(file_path, "fasta"))

# Function to annotate features in a sequence
def annotate_features(sequence, features):
    '''
    Annotate genomic features on the sequence.
    :param sequence: Sequence object.
    :param features: List of features to annotate.
    :return: Annotated sequence with features.
    '''
    annotated_seq = sequence
    for feature in features:
        feature_location = FeatureLocation(feature['start'],
            ↪ feature['end'])
```

```python
        seq_feature = SeqFeature(location=feature_location,
        ↪  type=feature['type'],
        ↪  qualifiers=feature.get('qualifiers', {}))
        annotated_seq.features.append(seq_feature)
    return annotated_seq

# Function to write annotated sequence to a file
def write_annotated_sequence(annotated_seq, output_file):
    '''
    Write the annotated sequence to a GenBank file.
    :param annotated_seq: Annotated sequence object.
    :param output_file: Path to the output file.
    '''
    with open(output_file, "w") as out_handle:
        SeqIO.write(annotated_seq, out_handle, "genbank")

# Sample usage workflow
fasta_file = "example.fasta"
genomic_features = [
    {'start': 100, 'end': 200, 'type': 'gene', 'qualifiers':
    ↪  {'gene': 'example1'}},
    {'start': 300, 'end': 400, 'type': 'exon', 'qualifiers':
    ↪  {'gene': 'example1', 'exon_number': '1'}}
]

# Parse sequences
sequences = parse_fasta(fasta_file)

# Annotate the first sequence with features
if sequences:
    annotated_sequence = annotate_features(sequences[0],
    ↪  genomic_features)
    output_path = "annotated_sequence.gb"
    write_annotated_sequence(annotated_sequence, output_path)
    print(f"Annotated sequence with features saved to
    ↪  {output_path}")
else:
    print("No sequences found to annotate.")
```

This code snippet demonstrates a simple pipeline for genomic feature annotation:

- `parse_fasta` function reads a FASTA file and extracts sequence records using `Biopython`.

- `annotate_features` adds feature annotations to a genomic sequence with specified locations and types.

- `write_annotated_sequence` exports the annotated sequence to a GenBank format file.

The example workflow processes sequence data from a FASTA file, applies annotations, and writes the results to an output file, demonstrating the end-to-end annotation workflow.

Chapter 47

Functional Annotation of Genes and Proteins

Below is a Python code snippet that exemplifies how functional annotation of genes and proteins can be accomplished using available bioinformatics data sources and common computational approaches.

```python
import numpy as np
from sklearn.ensemble import RandomForestClassifier
from sklearn.model_selection import train_test_split
from sklearn.metrics import accuracy_score
from Bio import SeqIO

def extract_features(sequence):
    '''
    Feature extraction from a protein or gene sequence.
    :param sequence: The sequence to extract features from.
    :return: A feature vector.
    '''
    # Example features: length and composition
    length = len(sequence)
    a_count = sequence.count('A') / length
    c_count = sequence.count('C') / length
    g_count = sequence.count('G') / length
    t_count = sequence.count('T') / length

    return [length, a_count, c_count, g_count, t_count]

def load_data(fasta_file, label_file):
    '''
    Loads sequences and their corresponding labels.
    :param fasta_file: Path to the FASTA file containing sequences.
```

```python
    :param label_file: Path to the file containing sequence labels.
    :return: Feature matrix and label array.
    '''
    # Load sequences
    sequences = []
    for record in SeqIO.parse(fasta_file, "fasta"):
        features = extract_features(str(record.seq))
        sequences.append(features)

    # Load labels
    with open(label_file, 'r') as lf:
        labels = [line.strip() for line in lf]

    return np.array(sequences), np.array(labels)

def train_annotation_model(X, y):
    '''
    Train a classification model to annotate sequences.
    :param X: Feature matrix.
    :param y: Labels.
    :return: Trained model.
    '''
    X_train, X_test, y_train, y_test = train_test_split(X, y,
     ↪ test_size=0.2, random_state=42)
    model = RandomForestClassifier(n_estimators=100,
     ↪ random_state=42)
    model.fit(X_train, y_train)
    predictions = model.predict(X_test)
    print("Accuracy:", accuracy_score(y_test, predictions))
    return model

def annotate_sequence(sequence, model):
    '''
    Annotate a single sequence using the trained model.
    :param sequence: The sequence to annotate.
    :param model: Trained model.
    :return: Predicted annotation.
    '''
    features = np.array(extract_features(sequence)).reshape(1, -1)
    return model.predict(features)[0]

# File paths
fasta_file = 'sequences.fasta'
label_file = 'labels.txt'

# Load data
X, y = load_data(fasta_file, label_file)

# Train model
annotation_model = train_annotation_model(X, y)

# Example annotation
test_sequence = "ATGCATGCA"
```

```
prediction = annotate_sequence(test_sequence, annotation_model)
print("The predicted function for the test sequence is:",
 ↪  prediction)
```

This code snippet aspires to demonstrate the essential components required for functional annotation of genetic sequences:

- `extract_features` function serves to distill important features from genetic sequences, such as length and nucleotide composition.

- `load_data` reads and processes data files, extracting sequence information and corresponding labels necessary for training the model.

- `train_annotation_model` trains a machine learning model to learn the relationship between sequence features and their functional annotations, utilizing a Random Forest classifier.

- `annotate_sequence` applies the pre-trained classification model to new sequences, offering predictions on their functional roles.

The sample block of code demonstrates loading sequence data, training a model, and performing a functional annotation on a new sequence.

Chapter 48

Pathway Enrichment Analysis

Below is a Python code snippet that encompasses the core computational elements for identifying significantly enriched pathways in omics datasets using an enrichment analysis approach like Gene Set Enrichment Analysis (GSEA). This method involves scoring a predefined gene set and assessing its enrichment compared to random gene sets.

```
import numpy as np
from scipy.stats import norm, rankdata
from collections import defaultdict

def compute_enrichment_score(gene_expression, gene_set,
 ↪  permute=False):
    '''
    Compute the enrichment score for a gene set based on
    ↪  differential expression data.
    :param gene_expression: Dictionary of gene names and their
    ↪  expression scores.
    :param gene_set: Set of genes to compute enrichment for.
    :param permute: Boolean to indicate if permutation of scores is
    ↪  needed (for significance testing).
    :return: Enrichment Score (ES).
    '''
    # Rank genes by expression score
    ranked_genes = rankdata([-gene_expression[gene] for gene in
    ↪  gene_expression])
    N = len(gene_expression)
    Nh = len(gene_set)
    # Indicator array for the presence of gene in the gene set
```

```python
    hits = np.array([1 if gene in gene_set else 0 for gene in
    ↪   gene_expression])
    # Rank normalization factors
    norm_factor = np.sum(hits) / Nh
    # Weighted enrichment calculation
    running_sum = np.cumsum(hits - (1/N))
    es = running_sum[np.argmax(np.abs(running_sum))] / norm_factor
    if permute:
        np.random.shuffle(running_sum)
    return es

def permutation_test(gene_expression, gene_set,
↪   n_permutations=1000):
    '''
    Conduct permutation test to calculate the significance of the
    ↪   enrichment score.
    :param gene_expression: Dictionary of gene names and their
    ↪   expression scores.
    :param gene_set: Set of genes to compute enrichment for.
    :param n_permutations: Number of permutations for significance
    ↪   testing.
    :return: p-value.
    '''
    observed_es = compute_enrichment_score(gene_expression,
    ↪   gene_set)
    permuted_es = [compute_enrichment_score(
        gene_expression, gene_set, permute=True) for _ in
        ↪   range(n_permutations)]
    return np.mean(np.abs(permuted_es) >= np.abs(observed_es))

# Hypothetical gene expression scores and a sample gene set
gene_scores = {'gene_' + str(i): np.random.normal() for i in
↪   range(100)}
sample_gene_set = {'gene_' + str(i) for i in
↪   np.random.choice(range(100), 20, replace=False)}

# Enrichment score of the sample gene set
es = compute_enrichment_score(gene_scores, sample_gene_set)
p_value = permutation_test(gene_scores, sample_gene_set)

print(f"Enrichment Score: {es}")
print(f"p-value from permutation test: {p_value}")
```

This code defines several key functions necessary for the implementation of pathway enrichment analysis:

- `compute_enrichment_score` calculates the enrichment score for a specific gene set using cumulative sums and ranking.

- `permutation_test` assesses the significance of the computed enrichment score through permutation testing.

The final block of code provides an example with hypothetical data, computing the enrichment score and determining the significance of the result using a permutation test.

Chapter 49

Comparative Genomics Algorithms

Below is a Python code snippet that implements core computational elements for comparing genomes to identify conserved and divergent regions, utilizing sequence alignment and comparative analysis techniques.

```python
from Bio import pairwise2
from Bio.pairwise2 import format_alignment
from Bio.Seq import Seq
import numpy as np

def global_align_sequences(seq1, seq2):
    '''
    Perform global alignment between two sequences.
    :param seq1: First DNA sequence.
    :param seq2: Second DNA sequence.
    :return: Alignment object with score and alignments.
    '''
    alignments = pairwise2.align.globalxx(seq1, seq2)
    return alignments

def calculate_conserved_divergent_regions(alignments):
    '''
    Analyze alignments to identify conserved and divergent regions.
    :param alignments: List of alignment tuples from global
    ↪ alignment.
    :return: Dictionary with conserved and divergent segments of the
    ↪ sequences.
    '''
    conserved = []
    divergent = []
```

```python
        seq1_aligned, seq2_aligned, score, start, end = alignments[0]

        for i, (a, b) in enumerate(zip(seq1_aligned, seq2_aligned)):
            if a == b:
                conserved.append((i, a))
            else:
                divergent.append((i, (a, b)))

        return {
            'conserved': conserved,
            'divergent': divergent
        }

def summarize_alignment_results(alignments, conserved_divergent):
    '''
    Print the alignment results and a summary of conserved and
    ↪    divergent regions.
    :param alignments: List of alignment objects.
    :param conserved_divergent: Dictionary of conserved and
    ↪    divergent regions.
    '''
    for alignment in alignments:
        print(format_alignment(*alignment))

    print("\nConserved Regions:")
    print(conserved_divergent['conserved'])

    print("\nDivergent Regions:")
    print(conserved_divergent['divergent'])

# Example sequences
seq1 = Seq("ATGCTGACCTAG")
seq2 = Seq("ATGCGTAACTAG")

# Perform the alignment
alignments = global_align_sequences(seq1, seq2)

# Calculate conserved and divergent regions
conserved_divergent =
↪    calculate_conserved_divergent_regions(alignments)

# Output the results
summarize_alignment_results(alignments, conserved_divergent)
```

This code snippet includes several key functions to analyze and compare genomic sequences:

- `global_align_sequences` uses the `Bio.pairwise2` module to perform global sequence alignment considering sequence similarity.

- `calculate_conserved_divergent_regions` identifies which parts of the alignment are conserved (identical in both sequences) and which are divergent (differing between sequences).

- `summarize_alignment_results` presents the alignments and summarizes conserved and divergent regions for interpretation.

The example given showcases these functions with two sample DNA sequences, providing both alignment and region analysis results.

Chapter 50

Ortholog and Paralog Identification

Below is a Python code snippet that demonstrates an algorithm to distinguish between orthologous and paralogous genes in genomic data using sequence alignment and phylogenetic tree analysis.

```
from Bio import Phylo, AlignIO
from Bio.Align.Applications import ClustalwCommandline
from Bio.Phylo.TreeConstruction import DistanceCalculator,
    ↪ DistanceTreeConstructor

def run_clustalw(input_file):
    '''
    Perform multiple sequence alignment using ClustalW.
    :param input_file: Path to the file containing sequences.
    :return: Path to the alignment file.
    '''
    clustalw_cline = ClustalwCommandline("clustalw2",
    ↪ infile=input_file)
    std_output, std_error = clustalw_cline()
    return input_file.replace('.fasta', '.aln')

def construct_phylogenetic_tree(aligned_file):
    '''
    Construct a phylogenetic tree from an aligned file.
    :param aligned_file: Path to the alignment file.
    :return: Phylogenetic tree object.
    '''
    alignment = AlignIO.read(aligned_file, 'clustal')
    calculator = DistanceCalculator('identity')
    dm = calculator.get_distance(alignment)
    constructor = DistanceTreeConstructor()
```

```
        tree = constructor.upgma(dm)
        return tree

def distinguish_orthologs_paralogs(tree):
    '''
    Analyze a phylogenetic tree to distinguish orthologous and
    ↪   paralogous genes.
    :param tree: Phylogenetic tree object.
    :return: Dictionay indicating orthologous and paralogous groups.
    '''
    clades = {}
    for clade in tree.find_clades():
        species_list = [term.name for term in clade.get_terminals()]
        if len(species_list) > 1:
            relationship = "Orthologs" if all(species in
            ↪   some_species_criteria for species in species_list)
            ↪   else "Paralogs"
            clades[clade.name] = {"relationship": relationship,
            ↪   "species": species_list}
    return clades

def main():
    input_file = 'genes.fasta'
    aligned_file = run_clustalw(input_file)
    tree = construct_phylogenetic_tree(aligned_file)
    ortholog_paralog_dict = distinguish_orthologs_paralogs(tree)

    for clade, info in ortholog_paralog_dict.items():
        print(f"Clade: {clade}, Relationship:
        ↪   {info['relationship']}, Species: {info['species']}")

if __name__ == "__main__":
    main()
```

This code defines the necessary components to distinguish between orthologous and paralogous genes:

- `run_clustalw` executes multiple sequence alignment using ClustalW for the given input file.

- `construct_phylogenetic_tree` creates a phylogenetic tree from the aligned sequence data.

- `distinguish_orthologs_paralogs` processes the tree to identify clades as orthologous or paralogous.

- `main` function ties together initialization, processing, and output of ortholog/paralog relationships.

By examining genetic sequences and their evolutionary relationships through phylogenetic trees, this code provides a clear methodology for classifying genes into orthologs and paralogs, essential for understanding gene evolution and function.

Chapter 51

Structural Variation Detection Algorithms

Below is a Python code snippet that encompasses the core computational elements for detecting structural variations such as inversions, translocations, and copy-number variations (CNVs) in genomic sequences. The algorithm utilizes paired-end sequencing reads for identifying and classifying these variations.

```python
import numpy as np
import pandas as pd
from scipy.stats import norm
from collections import defaultdict

def detect_structural_variations(read_pairs):
    '''
    Detect structural variations (SVs) using paired-end reads data.
    :param read_pairs: List of tuples representing paired-end reads.
     Each tuple has (read1_start, read2_start).
    :return: Dictionary with identified SVs categorized as
     inversion, translocation, CNV, etc.
    '''
    sv_dict = defaultdict(list)
    for read1, read2 in read_pairs:
        if read1 > read2:   # Suspected inversion
            sv_dict['inversion'].append((read2, read1))
        else:
            # Calculate the distance
            distance = read2 - read1
            # Basic heuristic for detecting structural variations
            if distance > 10000:   # Arbitrary threshold for CNV or
             translocation
                if np.random.rand() > 0.5:
```

```python
                sv_dict['translocation'].append((read1, read2))
            else:
                sv_dict['cnv'].append((read1, read2))

    return sv_dict

def simulate_read_pairs(num_pairs=1000):
    '''
    Simulates paired-end reads.
    :param num_pairs: Number of paired-end reads to simulate.
    :return: List of tuples with simulated read start positions.
    '''
    return [(np.random.randint(0, 1000000), np.random.randint(0,
        1000000)) for _ in range(num_pairs)]

def analyze_sv_results(sv_results):
    '''
    Analyze structural variation results to estimate occurrences and
        patterns.
    :param sv_results: Dictionary of structural variations.
    :return: Dataframe summarizing the structural variation
        analysis.
    '''
    analysis_dict = {'SV_Type': [], 'Occurrences': []}
    for sv_type, sv_list in sv_results.items():
        analysis_dict['SV_Type'].append(sv_type)
        analysis_dict['Occurrences'].append(len(sv_list))
    return pd.DataFrame(analysis_dict)

# Simulate read pairs
simulated_reads = simulate_read_pairs()

# Detect structural variations
sv_results = detect_structural_variations(simulated_reads)

# Analyze structural variation patterns
sv_analysis = analyze_sv_results(sv_results)

print(sv_analysis)
```

This code defines the following key computational components for detecting and analyzing structural variations:

- `detect_structural_variations` examines paired-end read data to identify potential inversions, translocations, and CNVs based on read positions and distances.

- `simulate_read_pairs` generates random paired-end read positions for testing and demonstration purposes.

- `analyze_sv_results` processes the results from structural variation detection, summarizing SV occurrences for analysis.

This concludes a comprehensive implementation framework for detecting large-scale genomic changes with example data simulation and result interpretation.

Chapter 52

Gene Fusion Detection in Cancer Genomics

Below is a Python code snippet that demonstrates how one might develop an algorithm for identifying gene fusions and their implications in cancer genomics, involving sequence alignment, fusion prediction, and annotation processes.

```
import re
import numpy as np
import pandas as pd

def load_gene_sequences(file_path):
    '''
    Load gene sequences from a FASTA file.
    :param file_path: Path to the FASTA file.
    :return: Dictionary with gene ids as keys and sequences as
    ↪ values.
    '''
    gene_sequences = {}
    with open(file_path, 'r') as file:
        gene_id = ""
        sequence = ""
        for line in file:
            line = line.strip()
            if line.startswith('>'):
                if gene_id:
                    gene_sequences[gene_id] = sequence
                gene_id = line[1:]  # remove the '>'
                sequence = ""
            else:
                sequence += line
        if gene_id:
```

```python
            gene_sequences[gene_id] = sequence
    return gene_sequences

def detect_fusion_genes(sequence_database, minimum_overlap=20):
    '''
    Detect potential gene fusions based on sequence overlaps.
    :param sequence_database: Dictionary of gene sequences.
    :param minimum_overlap: Minimum overlap in base pairs to
    ↪  consider a fusion.
    :return: List of tuples with gene pairs that form potential
    ↪  fusions.
    '''
    fusion_candidates = []
    for gene_id1, seq1 in sequence_database.items():
        for gene_id2, seq2 in sequence_database.items():
            if gene_id1 != gene_id2:
                if seq1[-minimum_overlap:] ==
                ↪   seq2[:minimum_overlap]:
                    fusion_candidates.append((gene_id1, gene_id2))
    return fusion_candidates

def annotate_fusion_gene(fusion_pair, gene_annotations):
    '''
    Annotate identified fusion genes with additional biological
    ↪  information.
    :param fusion_pair: Tuple of gene ids representing a fusion.
    :param gene_annotations: Dataframe containing gene annotations.
    :return: Annotation details for the fusion gene.
    '''
    annotation1 = gene_annotations.loc[gene_annotations['GeneID'] ==
    ↪   fusion_pair[0]]
    annotation2 = gene_annotations.loc[gene_annotations['GeneID'] ==
    ↪   fusion_pair[1]]
    return pd.concat([annotation1, annotation2])

# Example of loading gene annotations
def load_gene_annotations(filepath):
    '''
    Load gene annotations from a CSV file.
    :param filepath: Path to the CSV file.
    :return: Dataframe with gene annotations.
    '''
    return pd.read_csv(filepath)

# Conduct Fusion gene analysis
genes_sequences = load_gene_sequences("gene_sequences.fasta")
gene_annotations = load_gene_annotations("gene_annotations.csv")
potential_fusions = detect_fusion_genes(genes_sequences)

# Annotate each detected fusion gene
for fusion in potential_fusions:
    fusion_annotation = annotate_fusion_gene(fusion,
    ↪   gene_annotations)
```

```
print(f"Fusion: {fusion[0]} + {fusion[1]}")
print(fusion_annotation)
```

This code involves several critical components for detecting and analyzing gene fusions:

- `load_gene_sequences` loads gene sequences from a FASTA file, storing them in a dictionary.

- `detect_fusion_genes` identifies potential gene fusions by checking sequence overlaps between different genes.

- `annotate_fusion_gene` uses pre-existing annotations to provide additional context about the detected fusions.

- `load_gene_annotations` reads gene annotations from a CSV file for subsequent use in fusion annotation.

This framework serves as a starting point for more extensive gene fusion analysis in cancer genomics, where accurate detection and annotation are crucial for understanding gene interactions and potential implications in cancer studies.

Chapter 53

Protein-Protein Interaction Networks

Below is a Python code snippet that illustrates the computational construction and analysis of protein-protein interaction networks, including the building of the interaction network, network analysis, and visualization using libraries such as NetworkX and Matplotlib.

```python
import networkx as nx
import matplotlib.pyplot as plt
from typing import List, Tuple

def build_interaction_network(protein_pairs: List[Tuple[str, str]])
    -> nx.Graph:
    '''
    Constructs a protein interaction network from a list of protein
        pairs.
    :param protein_pairs: List of tuples representing protein
        interactions.
    :return: A NetworkX graph representing the protein interaction
        network.
    '''
    G = nx.Graph()
    G.add_edges_from(protein_pairs)
    return G

def analyze_network(G: nx.Graph) -> dict:
    '''
    Analyzes the protein interaction network.
    :param G: Protein interaction network graph.
    :return: Dictionary of network metrics such as degree
        centrality, clustering coefficient, etc.
    '''
```

```python
    analysis = {
        'degree_centrality': nx.degree_centrality(G),
        'clustering_coefficient': nx.average_clustering(G),
        'number_of_nodes': G.number_of_nodes(),
        'number_of_edges': G.number_of_edges(),
        'average_degree_connectivity':
            nx.average_degree_connectivity(G),
    }
    return analysis

def visualize_network(G: nx.Graph, title: str = 'Protein Interaction
    Network'):
    '''
    Visualizes the protein interaction network using Matplotlib.
    :param G: Protein interaction network graph.
    :param title: Title for the plot.
    '''
    pos = nx.spring_layout(G)
    plt.figure(figsize=(10, 10))
    nx.draw(G, pos, node_size=700, with_labels=True,
        node_color="skyblue", font_size=15, edge_color="black")
    plt.title(title)
    plt.show()

# Example usage with dummy protein interaction data
protein_interactions = [
    ('Protein1', 'Protein2'),
    ('Protein2', 'Protein3'),
    ('Protein3', 'Protein4'),
    ('Protein4', 'Protein5'),
    ('Protein5', 'Protein1')
]

# Build and analyze the network
G = build_interaction_network(protein_interactions)
network_analysis = analyze_network(G)

# Visualize the network
visualize_network(G)

# Print analysis results
print("Network Analysis Results:")
for metric, value in network_analysis.items():
    print(f"{metric}: {value}")
```

This code defines several key functions necessary for constructing and analyzing protein-protein interaction networks:

- `build_interaction_network` creates a graph from the provided list of protein interaction pairs.

- `analyze_network` calculates various network metrics such as degree centrality and clustering coefficient to assess network properties.

- `visualize_network` uses Matplotlib to render a visual representation of the interaction network, aiding in qualitative analysis.

- Example usage is provided at the end to demonstrate building, analyzing, and visualizing a simple network.

The code uses dummy protein interaction data to exhibit the functions' capabilities, printing out network metrics and displaying the network graph.

Chapter 54

Network Topology Analysis in Systems Biology

Below is a Python code snippet that explores the structural properties of biological networks to infer functional insights, emphasizing network topology analysis through graph-related operations and calculations.

```python
import networkx as nx
import matplotlib.pyplot as plt

def analyze_network_topology(graph):
    '''
    Analyze the topology of a biological network.
    :param graph: NetworkX graph object representing the biological
    ↪   network.
    :return: Various network topology metrics.
    '''
    metrics = {}

    # Calculate degree centrality
    metrics['degree_centrality'] = nx.degree_centrality(graph)

    # Calculate betweenness centrality
    metrics['betweenness_centrality'] =
    ↪   nx.betweenness_centrality(graph)

    # Calculate closeness centrality
    metrics['closeness_centrality'] = nx.closeness_centrality(graph)
```

```python
        # Calculate clustering coefficient
        metrics['clustering_coefficient'] = nx.clustering(graph)

        # Determine number of connected components
        metrics['connected_components'] =
        ↪    nx.number_connected_components(graph)

        # Average shortest path length (for connected graphs)
        if nx.is_connected(graph):
            metrics['average_shortest_path_length'] =
            ↪    nx.average_shortest_path_length(graph)
        else:
            metrics['average_shortest_path_length'] = None

        return metrics

    def plot_degree_distribution(graph):
        '''
        Plot the degree distribution of the network.
        :param graph: NetworkX graph object representing the biological
        ↪    network.
        '''
        degree_sequence = sorted([d for n, d in graph.degree()],
        ↪    reverse=True)
        plt.figure()
        plt.bar(*np.unique(degree_sequence, return_counts=True))
        plt.title("Degree Distribution")
        plt.xlabel("Degree")
        plt.ylabel("Frequency")
        plt.show()

    # Example of creating a biological network graph
    G = nx.erdos_renyi_graph(n=100, p=0.05)

    # Analyze network topology
    topology_metrics = analyze_network_topology(G)
    print("Network Topology Metrics:", topology_metrics)

    # Plot degree distribution
    plot_degree_distribution(G)
```

This code provides a comprehensive approach to analyzing biological network topology through several key functions:

- `analyze_network_topology` computes various centrality measures and clustering coefficients to uncover the network's structural characteristics.

- `plot_degree_distribution` visualizes the frequency of node degrees, helping to understand the distribution patterns within the network.

The final block of code initializes a random Erdős–Rényi graph as an example of a biological network, analyzes its topology, and visualizes its degree distribution using the implemented functions.

Chapter 55

Metagenomic Data Analysis Algorithms

Below is a Python code snippet that showcases the implementation for analyzing genetic material recovered directly from environmental samples. The process involves loading metagenomic sequence data, performing quality filtering, and using BLAST for taxonomic classification.

```python
import os
import subprocess
import pandas as pd
from Bio import SeqIO
from Bio.Blast import NCBIWWW, NCBIXML

def run_fastqc(filepath):
    '''
    Perform quality control using FastQC.
    :param filepath: Path to input FASTQ file.
    :return: Report file path.
    '''
    command = ["fastqc", filepath]
    subprocess.run(command)
    return filepath.replace('.fastq', '_fastqc.html')

def filter_low_quality_sequences(input_file, output_file,
    min_quality=30):
    '''
    Filter out low-quality sequences from FASTQ file.
    :param input_file: Input FASTQ file path.
    :param output_file: Output filtered FASTQ file path.
    :param min_quality: Minimum quality threshold for filtering.
```

```python
    '''
    with open(output_file, 'w') as out_handle:
        for record in SeqIO.parse(input_file, "fastq"):
            if min(record.letter_annotations["phred_quality"]) >=
            ↪ min_quality:
                SeqIO.write(record, out_handle, "fastq")

def blast_sequence(sequence):
    '''
    Run BLAST search on a given DNA sequence.
    :param sequence: DNA sequence string.
    :return: BLAST search result.
    '''
    result = NCBIWWW.qblast("blastn", "nt", sequence)
    return NCBIXML.read(result)

def parse_blast_results(blast_record):
    '''
    Extract significant hits from BLAST results.
    :param blast_record: BLAST record object.
    :return: List of relevant taxonomic IDs and descriptions.
    '''
    significant_hits = []
    for alignment in blast_record.alignments:
        for hsp in alignment.hsps:
            if hsp.expect < 0.01:  # Threshold for significant hit
                hit_info = {
                    'title': alignment.title,
                    'length': alignment.length,
                    'e-value': hsp.expect
                }
                significant_hits.append(hit_info)
    return significant_hits

# Paths to input/output files
input_fastq = 'environmental_sample.fastq'
filtered_fastq = 'filtered_sample.fastq'
blast_output_csv = 'blast_results.csv'

# Step 1: Run FastQC
qc_report = run_fastqc(input_fastq)
print(f"FastQC report generated: {qc_report}")

# Step 2: Filter out low-quality sequences
filter_low_quality_sequences(input_fastq, filtered_fastq)
print(f"Filtered sequences saved to: {filtered_fastq}")

# Step 3: Perform BLAST for taxonomic classification
all_results = []
for record in SeqIO.parse(filtered_fastq, "fastq"):
    blast_record = blast_sequence(str(record.seq))
    hits = parse_blast_results(blast_record)
    all_results.extend(hits)
```

```
# Step 4: Save results to CSV
df = pd.DataFrame(all_results)
df.to_csv(blast_output_csv, index=False)
print(f"BLAST results saved to: {blast_output_csv}")
```

This code snippet details a comprehensive approach to processing and analyzing metagenomic samples:

- `run_fastqc` function runs the FastQC tool to assess sequence quality.

- `filter_low_quality_sequences` removes sequences below a certain quality threshold.

- `blast_sequence` performs a BLAST search for taxonomic classification.

- `parse_blast_results` extracts significant hits from BLAST results.

The script showcases the integration of quality control and taxonomic analysis using sequence data from environmental samples, saving results into a convenient CSV format.

Chapter 56

Taxonomic Classification in Metagenomics

Below is a Python code snippet that provides a computational approach for classifying and identifying organisms within metagenomic datasets using a combination of preprocessing, sequence alignment, and machine learning techniques.

```
import numpy as np
import pandas as pd
from sklearn.preprocessing import StandardScaler
from sklearn.decomposition import PCA
from sklearn.ensemble import RandomForestClassifier

def preprocess_metagenomic_data(sequence_data):
    '''
    Preprocess input metagenomic sequence data.
    :param sequence_data: Raw nucleotide sequence data.
    :return: Processed and normalized feature matrix.
    '''
    # Example dummy encoding; in practice, use actual sequence
    ↪ transformations
    feature_matrix = np.random.rand(len(sequence_data), 100)  #
    ↪ dummy dimensionality
    scaler = StandardScaler()
    normalized_data = scaler.fit_transform(feature_matrix)
    return normalized_data

def perform_pca(normalized_data, n_components=10):
    '''
```

```
    Perform PCA to reduce dimensionality of the metagenomic data.
    :param normalized_data: Feature matrix of normalized data.
    :param n_components: Number of PCA components to retain.
    :return: Reduced dimension feature matrix.
    '''
    pca = PCA(n_components=n_components)
    reduced_data = pca.fit_transform(normalized_data)
    return reduced_data

def align_sequences_to_database(sequence_data, database):
    '''
    Align sequences against a known database to gather feature
    ↪ alignments.
    :param sequence_data: Input sequence data for alignment.
    :param database: Database to align against.
    :return: Alignment scores or features.
    '''
    # Placeholder for sequence alignment logic
    alignment_scores = np.random.rand(len(sequence_data),
    ↪ len(database))
    return alignment_scores

def classify_organisms(features, labels, test_data):
    '''
    Train a classifier and predict organism types in test samples.
    :param features: Training feature matrix.
    :param labels: Known labels for training data.
    :param test_data: Feature matrix for test data.
    :return: Predicted labels for test data.
    '''
    clf = RandomForestClassifier(n_estimators=100, random_state=42)
    clf.fit(features, labels)
    predictions = clf.predict(test_data)
    return predictions

# Example usage
sequence_data = ['AGCT', 'CGTA', 'TACG', 'GTAC']
database = ['AGCT', 'CGTA', 'TACG', 'GTAC', 'XXXX']  # Placeholder
↪ database sequences

# Preprocess data
normalized_data = preprocess_metagenomic_data(sequence_data)
reduced_data = perform_pca(normalized_data)

# Align to database
alignment_scores = align_sequences_to_database(sequence_data,
↪ database)

# Dummy labels for demonstration
labels = np.random.choice(['Species1', 'Species2'],
↪ len(sequence_data))

# Classify organisms
```

```
predictions = classify_organisms(reduced_data, labels, reduced_data)
print("Predicted Organisms:", predictions)
```

This code provides a comprehensive look into a pipeline for metagenomic classification:

- `preprocess_metagenomic_data` function normalizes raw sequence data using a feature matrix approach with standard scaling.

- `perform_pca` reduces feature dimensionality for more efficient classification.

- `align_sequences_to_database` is a placeholder for aligning sequences to a biological database, essential for identifying taxonomic features.

- `classify_organisms` exemplifies training a classification model to predict organism presence in metagenomic samples leveraging the `RandomForestClassifier`.

The provided code example demonstrates preprocessing, dimension reduction, sequence alignment, and subsequent classification with dummy data, showcasing the typical steps in handling metagenomic datasets.

Chapter 57

Functional Profiling of Microbial Communities

Below is a Python code snippet that encompasses the core computational elements for predicting the functional capabilities of microbial communities using functional profiling algorithms. This includes the prediction pipeline setup, implementation of key functions, and simulation of microbial interactions.

```python
import numpy as np
from sklearn.decomposition import PCA

def load_metagenomic_data(file_path):
    '''
    Load metagenomic data from a file.
    :param file_path: Path to the file containing metagenomic data.
    :return: A numpy array of the data.
    '''
    # For demonstration: simulate loading data
    data = np.random.rand(100, 50)  # 100 samples, 50 features
    return data

def functional_annotation(feature_array):
    '''
    Annotate features with functions.
    :param feature_array: Array of features for annotation.
    :return: Annotated functional capabilities.
    '''
    # Simulate annotation to functions
    functional_mapping = {feature: np.random.rand() for feature in
        range(feature_array.shape[1])}
    return functional_mapping
```

```
def predict_functional_capabilities(data, functional_mapping):
    '''
    Predict the functional capabilities based on the mapping and
    ↪ data.
    :param data: Metagenomic data array.
    :param functional_mapping: Mapping of features to functional
    ↪ capabilities.
    :return: Predicted functional capabilities for the community.
    '''
    capabilities = np.dot(data,
    ↪ np.array(list(functional_mapping.values())))
    return capabilities

def reduce_dimensions(data):
    '''
    Perform dimensionality reduction on the data.
    :param data: Original data array.
    :return: Reduced data array.
    '''
    pca = PCA(n_components=10)
    reduced_data = pca.fit_transform(data)
    return reduced_data

# Example pipeline execution
file_path = 'metagenomic_data.txt'
metagenomic_data = load_metagenomic_data(file_path)
functional_mapping = functional_annotation(metagenomic_data)
functional_capabilities =
↪ predict_functional_capabilities(metagenomic_data,
↪ functional_mapping)
reduced_data = reduce_dimensions(metagenomic_data)

print("Functional Capabilities Predicted:", functional_capabilities)
print("Dimensionality-Reduced Data Shape:", reduced_data.shape)
```

This code defines the fundamental functions necessary for predicting functional capabilities within microbial communities through functional profiling:

- `load_metagenomic_data` simulates the loading of metagenomic datasets from storage, providing raw data necessary for analysis.

- `functional_annotation` maps metagenomic features to their respective functional roles, a crucial step for assessing community capabilities.

- `predict_functional_capabilities` uses the annotation mapping to estimate the community's functional profile.

- `reduce_dimensions` applies Principal Component Analysis (PCA) for complexity reduction, enabling easier interpretation and analysis of the data.

The execution block demonstrates a simple pipeline of loading, annotating, predicting, and reducing dimensions for a mock dataset.

Chapter 58

Genome Reconstruction from Metagenomic Data

Below is a Python code snippet that encompasses the core computational elements for reconstructing individual genomes from complex metagenomic datasets using a basic assembly approach. This includes functions for k-mer generation, de Bruijn graph construction, and genome assembly.

```python
import itertools
from collections import defaultdict, deque

def kmer_generator(sequence, k):
    '''
    Generate k-mers from a given sequence.
    :param sequence: Input DNA sequence.
    :param k: Length of k-mer.
    :return: List of k-mers.
    '''
    return [sequence[i:i+k] for i in range(len(sequence) - k + 1)]

def build_debruijn_graph(kmers):
    '''
    Construct a de Bruijn graph from a list of k-mers.
    :param kmers: List of k-mers.
    :return: Adjacency list representing the graph.
    '''
    adj_list = defaultdict(list)
    for kmer in kmers:
```

```python
            prefix = kmer[:-1]
            suffix = kmer[1:]
            adj_list[prefix].append(suffix)
    return adj_list

def find_eulerian_path(graph):
    '''
    Find Eulerian path from a given de Bruijn graph.
    :param graph: de Bruijn graph's adjacency list.
    :return: Eulerian path as a list of nodes.
    '''
    in_degree = defaultdict(int)
    out_degree = defaultdict(int)

    for node in graph:
        out_degree[node] += len(graph[node])
        for adjacent in graph[node]:
            in_degree[adjacent] += 1

    start_node, end_node = None, None
    for node in set(in_degree) | set(out_degree):
        out = out_degree[node]
        inp = in_degree[node]
        if out > inp:
            start_node = node
        elif inp > out:
            end_node = node

    if not start_node:
        start_node = next(iter(graph))

    path = []
    stack = [start_node]
    while stack:
        u = stack[-1]
        if out_degree[u] == 0:
            path.append(stack.pop())
        else:
            v = graph[u].pop()
            out_degree[u] -= 1
            stack.append(v)
    return path[::-1]

def assemble_genome_from_path(eulerian_path):
    '''
    Assemble genome sequence from an Eulerian path.
    :param eulerian_path: List of nodes representing Eulerian path.
    :return: Reconstructed genome sequence.
    '''
    assembled_sequence = eulerian_path[0]
    for node in eulerian_path[1:]:
        assembled_sequence += node[-1]
    return assembled_sequence
```

```python
# Example usage with a mock DNA sequence
dna_sequence = "AACGTCGGTACCAAACGCTA"

# Length of k-mers
k = 3

# Generate k-mers
kmers = kmer_generator(dna_sequence, k)

# Build the de Bruijn graph
debruijn_graph = build_debruijn_graph(kmers)

# Find the Eulerian path
eulerian_path = find_eulerian_path(debruijn_graph)

# Assemble the genome
assembled_genome = assemble_genome_from_path(eulerian_path)

print("Assembled Genome:", assembled_genome)
```

This code defines several key functions necessary for genome reconstruction from metagenomic data:

- `kmer_generator` generates k-mers from the input DNA sequence, which are the basis for constructing the graph.

- `build_debruijn_graph` creates an adjacency list that represents the de Bruijn graph from the k-mers.

- `find_eulerian_path` determines the Eulerian path through the graph, which will enumerate all edges exactly once.

- `assemble_genome_from_path` reconstructs the genome sequence from the Eulerian path.

The final block of code demonstrates these functions with a simple DNA sequence, thus illustrating the general method of genome assembly from complex metagenomic datasets.

Chapter 59

Gene Ontology Enrichment Analysis

Below is a Python code snippet that applies ontology enrichment analysis for functional annotation and understanding of gene sets using statistical methods to determine enrichment of functional categories.

```
import pandas as pd
from scipy.stats import hypergeom

def load_gene_sets(gene_set_file):
    '''
    Load predefined gene sets from a file.
    :param gene_set_file: Filepath to predefined gene set.
    :return: Dictionary mapping of gene sets.
    '''
    gene_sets = {}
    with open(gene_set_file, 'r') as f:
        for line in f:
            line = line.strip().split("\t")
            gene_set_name = line[0]
            genes = line[1].split(",")
            gene_sets[gene_set_name] = set(genes)
    return gene_sets

def perform_ontology_enrichment(test_gene_set, gene_sets,
    ↪ total_genes_in_universe):
    '''
    Perform ontology enrichment analysis.
    :param test_gene_set: The gene set to be tested for enrichment.
    :param gene_sets: Predefined sets of genes associated with
        ↪ functional categories.
```

```
    :param total_genes_in_universe: Total number of genes in the
    ↪ universe.
    :return: DataFrame with enrichment results.
    '''
    results = []
    M = total_genes_in_universe
    n = len(test_gene_set)

    for category, genes in gene_sets.items():
        K = len(genes)
        x = len(test_gene_set.intersection(genes))
        # Hypergeometric test
        p_value = hypergeom.sf(x-1, M, K, n)
        results.append({
            "Category": category,
            "Overlap": x,
            "Category Size": K,
            "Gene Set Size": n,
            "P-Value": p_value
        })

    return pd.DataFrame(results).sort_values(by='P-Value')

# Example genes
test_gene_set = {'GeneA', 'GeneB', 'GeneC', 'GeneD'}

# Load predefined gene sets
gene_sets = load_gene_sets('gene_sets.txt')

# Perform enrichment analysis
enrichment_results = perform_ontology_enrichment(test_gene_set,
    ↪ gene_sets, total_genes_in_universe=20000)

# Output results
print(enrichment_results)
```

This code defines a process for conducting ontology enrichment analysis to functionally annotate gene sets:

- `load_gene_sets` function handles loading predefined gene sets which are typically stored in a text file format with each set grouped by functional categories.

- `perform_ontology_enrichment` conducts the enrichment analysis using the hypergeometric distribution to calculate the p-values indicating the significance of overlap between the test gene set and each functional category.

- The enrichment analysis produces a DataFrame that categorizes results by functional category, overlap count, category size, and p-value, sorted by statistical significance.

This implementation allows for a robust approach to discern which biological functions are most strongly represented within a given set of genes relative to a background universe of gene categories.

Chapter 60

Antibody Design Algorithms

Below is a Python code snippet that demonstrates computational methods for predicting and optimizing antibody sequences for therapeutics including sequence modeling, affinity prediction, and algorithmic optimization for therapeutic antibody design.

```python
import numpy as np
import random

def generate_antibody_sequences(length, num_sequences=1):
    '''
    Generate random antibody sequences of a given length.
    :param length: Length of each antibody sequence.
    :param num_sequences: Number of sequences to generate.
    :return: List of random antibody sequences.
    '''
    amino_acids = 'ACDEFGHIKLMNPQRSTVWY'
    return [''.join(random.choices(amino_acids, k=length)) for _ in
    ↪    range(num_sequences)]

def calculate_binding_affinity(sequence):
    '''
    Simulate the calculation of binding affinity for an antibody
    ↪    sequence.
    :param sequence: Antibody amino acid sequence.
    :return: Simulated binding affinity value.
    '''
    # In practice, this would be replaced with an actual model or
    ↪    experimental data
    affinity = sum(ord(char) for char in sequence) % 100
    return affinity
```

```python
def optimize_sequence(sequence):
    '''
    Optimize an antibody sequence for better binding affinity.
    :param sequence: Original antibody sequence.
    :return: Optimized antibody sequence.
    '''
    best_sequence = sequence
    best_affinity = calculate_binding_affinity(sequence)

    # Simulated annealing or other optimization algorithms can be
    #   used here
    for _ in range(1000):
        # Randomly modify the sequence
        new_sequence = list(sequence)
        new_sequence[random.randint(0, len(sequence) - 1)] = 
            random.choice('ACDEFGHIKLMNPQRSTVWY')
        new_sequence = ''.join(new_sequence)
        new_affinity = calculate_binding_affinity(new_sequence)
        # If new sequence is better, adopt it
        if new_affinity > best_affinity:
            best_affinity = new_affinity
            best_sequence = new_sequence

    return best_sequence

def main():
    '''
    Main function to generate, evaluate, and optimize antibody
        sequences.
    '''
    # Generate initial antibody sequences
    initial_sequences = generate_antibody_sequences(10, 5)

    for sequence in initial_sequences:
        print(f"Original Sequence: {sequence}, Binding Affinity:
            {calculate_binding_affinity(sequence)}")
        optimized_sequence = optimize_sequence(sequence)
        print(f"Optimized Sequence: {optimized_sequence}, Optimized
            Binding Affinity:
            {calculate_binding_affinity(optimized_sequence)}\n")

if __name__ == "__main__":
    main()
```

This code defines several key functions for working with antibody sequences:

- `generate_antibody_sequences` creates random sequences of amino acids, simulating potential antibodies.

- `calculate_binding_affinity` estimates the affinity of a given sequence for binding, serving as a placeholder for a real experimental or computational model.

- `optimize_sequence` attempts to improve the binding affinity of an antibody sequence using a simple optimization method, which could be replaced by more sophisticated algorithms like simulated annealing or genetic algorithms.

The `main` function demonstrates the generation and optimization process, illustrating before-and-after sequence affinities, emphasizing computational strategies for designing therapeutic antibodies.

Chapter 61

Computational Vaccine Design

Below is a Python code snippet for designing vaccines using reverse vaccinology and epitope prediction. This code covers the essential steps including processing antigen sequences, predicting potential epitopes, and validating them against known databases.

```python
import numpy as np
from Bio import SeqIO
from Bio.SeqUtils.ProtParam import ProteinAnalysis
from sklearn.ensemble import RandomForestClassifier
import pandas as pd

def load_antigen_sequences(file_path):
    '''
    Load antigen sequences from a FASTA file.
    :param file_path: Path to the FASTA file.
    :return: List of sequences.
    '''
    sequences = [str(record.seq) for record in
        SeqIO.parse(file_path, "fasta")]
    return sequences

def predict_epitopes(sequence, threshold=1.0):
    '''
    Predict potential epitopes from an antigen sequence.
    :param sequence: Antigen sequence.
    :param threshold: Hydrophilicity threshold for epitope
        selection.
    :return: List of predicted epitopes.
    '''
    analysis = ProteinAnalysis(sequence)
```

```python
    hydrophilicity = analysis.protein_scale(window=7,
        param_dict={"KD": {"ALA": 1.8, "ARG": -4.5, "ASN": -3.5,
        "ASP": -3.5}})
    epitopes = []
    for i, value in enumerate(hydrophilicity):
        if value > threshold:
            epitopes.append(sequence[i:i+7])
    return epitopes

def validate_epitopes(epitopes, database):
    '''
    Validate predicted epitopes against known epitope database.
    :param epitopes: List of predicted epitopes.
    :param database: DataFrame of known epitopes.
    :return: DataFrame of validated epitopes.
    '''
    validated = [epitope for epitope in epitopes if epitope in
        database["epitope"].values]
    return pd.DataFrame(validated, columns=["epitope"])

def train_epitope_classifier(epitope_data, labels):
    '''
    Train a classifier to predict epitope efficacy.
    :param epitope_data: Features of epitopes.
    :param labels: Known labels of efficacy.
    :return: Trained classifier.
    '''
    clf = RandomForestClassifier(n_estimators=100)
    clf.fit(epitope_data, labels)
    return clf

def main():
    file_path = "antigens.fasta"
    antigen_sequences = load_antigen_sequences(file_path)

    all_predicted_epitopes = []
    for seq in antigen_sequences:
        predicted = predict_epitopes(seq)
        all_predicted_epitopes.extend(predicted)

    known_epitopes_db = pd.DataFrame({'epitope': ['ATCG', 'GCTA',
        'TACG']})  # Example known epitopes
    validated_epitopes = validate_epitopes(all_predicted_epitopes,
        known_epitopes_db)

    # Example feature data and labels for training
    dummy_features = np.random.rand(len(validated_epitopes), 5)
    dummy_labels = np.random.randint(0, 2, len(validated_epitopes))

    clf = train_epitope_classifier(dummy_features, dummy_labels)

    print("Validated Epitopes:\n", validated_epitopes)
    print("Classifier Training Completed")
```

```
if __name__ == "__main__":
    main()
```

This code snippet includes essential functions for reverse vaccinology and epitope prediction:

- `load_antigen_sequences` function reads and loads antigen sequences from a FASTA file.
- `predict_epitopes` uses hydrophilicity calculations to predict potential epitopes within a sequence.
- `validate_epitopes` checks predicted epitopes against a database of known epitopes to filter out non-relevant sequences.
- `train_epitope_classifier` is a basic implementation to train a machine learning classifier to evaluate the efficacy of predicted epitopes.

The example demonstrates the pipeline from reading sequences to training a classifier, emphasizing the integration of computational and biological insights for vaccine design.

Chapter 62

In Silico Drug Discovery Platforms

Below is a Python code snippet that implements core computational elements used in virtual screening and molecular docking, providing a foundation for drug discovery processes. The code includes functions to perform molecular docking simulations and predict binding affinities.

```
import numpy as np
from scipy.spatial import distance_matrix

def generate_conformations(molecule, n_conformations=100):
    '''
    Generates a set of random conformations for a given molecule.
    :param molecule: A molecular structure in a standardized format
     (e.g., SMILES or PDB).
    :param n_conformations: Number of conformations to generate.
    :return: A list of conformations.
    '''
    # Placeholder for conformation generation logic
    conformations = []
    for _ in range(n_conformations):
        conformations.append(np.random.rand(3, len(molecule)))  #
         Random 3D points
    return conformations

def calculate_interaction_energy(protein, ligand):
    '''
    Calculates interaction energy between protein and ligand.
    :param protein: Protein structure as a set of 3D coordinates.
    :param ligand: Ligand structure as a set of 3D coordinates.
```

```python
    :return: Interaction energy.
    '''
    protein_coords = np.array(protein)
    ligand_coords = np.array(ligand)
    distances = distance_matrix(protein_coords, ligand_coords)

    # Simplified energy model: Sum of inverse distances
    energy = np.sum(1.0 / (distances + 1e-6))  # Avoid division by
    ↪  zero
    return energy

def dock_and_score(protein, ligand_library):
    '''
    Performs docking of a ligand library against a target protein to
    ↪  predict binding affinities.
    :param protein: Target protein structure.
    :param ligand_library: List of ligand molecules.
    :return: List of docking scores for each ligand.
    '''
    scores = []
    for ligand in ligand_library:
        conformations = generate_conformations(ligand)
        best_score = float('inf')
        for conformation in conformations:
            score = calculate_interaction_energy(protein,
            ↪  conformation)
            best_score = min(score, best_score)
        scores.append((ligand, best_score))
    return scores

# Example structures
protein_structure = np.random.rand(3, 15)  # Protein with 15 atoms
ligand_library = [
    "C(C(=O)O)c1ccccc1",  # Example SMILES for a small molecule
    "CCN(CC)CCCC(C)Nc1nc2n(c3ccccc3)c(=O)[nH]c2s1"  # Another
    ↪  example
]

# Convert SMILES or other formats to a suitable list of atom
↪  coordinates for docking
# protein_structure and ligand_library should already represent this
docking_results = dock_and_score(protein_structure, ligand_library)
for ligand, score in docking_results:
    print(f"Ligand: {ligand}, Docking Score: {score}")
```

This code defines key functions essential for conducting virtual screenings and molecular docking, which are crucial steps in drug discovery:

- `generate_conformations` generates random 3D structures

for each molecule, simulating different conformations that a ligand might adopt.

- `calculate_interaction_energy` computes the interaction energy between the protein and ligand, providing a rough estimate of the binding affinity.

- `dock_and_score` performs docking simulations by evaluating each conformation of the ligand and determining its potential binding score with the protein.

The snippet concludes with an example application, demonstrating the use of these functions to screen a small library of ligands against a protein target and predict their binding scores.

Chapter 63

Gene Therapy Vector Optimization

Below is a Python code snippet that demonstrates the implementation of key algorithms used in designing and optimizing viral vectors for gene therapy applications. This involves simulation of vector optimization, off-target effect prediction, and integration with relevant biological data for enhanced decision-making.

```python
import numpy as np
from scipy.optimize import minimize

def viral_vector_design(genes, target_cells,
    off_target_threshold=0.05):
    '''
    Optimize viral vector design to target specific genes in target
        cells.
    :param genes: List of genes to be targeted.
    :param target_cells: Characteristics of the target cells.
    :param off_target_threshold: Maximum allowable off-target
        binding.
    :return: Optimized vector features.
    '''
    def objective_function(vector_props):
        '''
        Objective function for optimization.
        :param vector_props: Properties of the viral vector.
        :return: Efficiency score (to be minimized).
        '''
        expression_levels = vector_expression_score(vector_props,
            genes)
        off_target_score = evaluate_off_target_effect(vector_props,
            target_cells)
```

```
            return -expression_levels + (off_target_threshold *
            ↪ off_target_score)

        # Initial guess for vector properties
        initial_guess = np.random.rand(len(genes))
        result = minimize(objective_function, initial_guess, bounds=[(0,
        ↪ 1)]*len(genes))
        return result.x

def vector_expression_score(vector_props, genes):
    '''
    Calculate the expression score for a given viral vector.
    :param vector_props: Properties of the vector.
    :param genes: List of target genes.
    :return: Computed expression score.
    '''
    return np.dot(vector_props, np.random.random(len(genes))) #
    ↪ Placeholder for real model

def evaluate_off_target_effect(vector_props, target_cells):
    '''
    Predict off-target effects of the viral vector.
    :param vector_props: Properties of the vector.
    :param target_cells: Characteristics of target cells.
    :return: Off-target score.
    '''
    return np.sum(vector_props *
    ↪ np.random.random(len(target_cells)))  # Placeholder for
    ↪ calculation

# Example of genes and target cell characteristics
genes = ['gene1', 'gene2', 'gene3']
target_cells = {'cell_type': 'hepatocyte', 'marker': 'APOB',
↪ 'expression_level': 0.8}

# Optimization process
optimized_vector = viral_vector_design(genes, target_cells)

print("Optimized Vector Properties:", optimized_vector)
```

This code defines the following functions key to viral vector optimization:

- `viral_vector_design` optimizes vector properties to maximize gene delivery efficiency while minimizing off-target effects. It uses `scipy.optimize.minimize` for parameter optimization.

- `vector_expression_score` evaluates how well the vector influences the expression of target genes, returning an expression score used in optimization.

- `evaluate_off_target_effect` predicts potential unintended interactions of the vector with non-target cells, generating an off-target score.

This framework enables the design of viral vectors tailored for specific gene therapy applications, fostering advancements in precision medicine. The example uses placeholder functions for scoring, which would be replaced by detailed biological models in practical applications.

Chapter 64

Algorithmic Approaches to Gene Synthesis

Below is a Python code snippet that encompasses the core computational elements of optimizing gene synthesis processes, including codon optimization.

```python
import random

def generate_random_dna_sequence(length):
    '''
    Generate a random DNA sequence of given length.
    :param length: Length of the DNA sequence to generate.
    :return: Random DNA sequence as a string.
    '''
    return ''.join(random.choices('ATCG', k=length))

def calculate_gc_content(dna_sequence):
    '''
    Calculate the GC content of a DNA sequence.
    :param dna_sequence: DNA sequence string.
    :return: GC content percentage.
    '''
    g_count = dna_sequence.count('G')
    c_count = dna_sequence.count('C')
    return (g_count + c_count) / len(dna_sequence) * 100

def codon_optimization(dna_sequence, preferred_codons):
    '''
    Optimize the codon usage of a DNA sequence.
```

```python
    :param dna_sequence: Input DNA sequence string.
    :param preferred_codons: Dictionary with amino acids as keys and
    ↪   preferred codons as values.
    :return: Codon-optimized DNA sequence.
    '''
    # A simple codon table for demonstration purposes
    codon_table = {
        'A': ['GCT', 'GCC', 'GCA', 'GCG'],
        'R': ['CGT', 'CGC', 'CGA', 'CGG', 'AGA', 'AGG'],
        'N': ['AAT', 'AAC'],
        'D': ['GAT', 'GAC'],
        'C': ['TGT', 'TGC'],
        # Add more amino acids and their codons as needed
    }

    optimized_sequence = ''
    for i in range(0, len(dna_sequence), 3):
        codon = dna_sequence[i:i+3]
        amino_acid = None

        # Look for the amino acid that the current codon codes for
        for aa, codons in codon_table.items():
            if codon in codons:
                amino_acid = aa
                break

        if amino_acid and amino_acid in preferred_codons:
            # Substitute with the preferred codon
            optimized_sequence += preferred_codons[amino_acid]
        else:
            # Keep the original codon if no substitution is found
            optimized_sequence += codon

    return optimized_sequence

# Example usage
original_sequence = generate_random_dna_sequence(30)
print("Original Sequence:", original_sequence)
print("Original GC Content:",
    ↪   calculate_gc_content(original_sequence))

preferred_codons = {
    'A': 'GCC',   # Example preferred codon for Alanine
    'R': 'CGT',   # Example preferred codon for Arginine
    'N': 'AAC',   # Example preferred codon for Asparagine
    'D': 'GAC',   # Example preferred codon for Aspartic acid
    'C': 'TGC',   # Example preferred codon for Cysteine
}

optimized_sequence = codon_optimization(original_sequence,
    ↪   preferred_codons)
print("Optimized Sequence:", optimized_sequence)
```

```
print("Optimized GC Content:",
    calculate_gc_content(optimized_sequence))
```

This code defines several key functions necessary for the optimization of gene synthesis processes:

- `generate_random_dna_sequence` generates a random DNA sequence given a specific length.

- `calculate_gc_content` calculates the GC content of a given DNA sequence.

- `codon_optimization` adjusts the input DNA sequence based on a provided set of preferred codons, contributing to optimized gene synthesis by favoring certain codons over others.

The final block of code demonstrates the process using an example sequence, printing both the original and optimized sequences along with their GC contents.

Chapter 65

Codon Usage Bias Analysis

Below is a Python code snippet focusing on analyzing and manipulating codon usage to enhance protein expression, including functions for calculating codon frequency, optimizing codon usage, and simulating protein expression levels.

```
import numpy as np
from collections import Counter

def codon_frequency(sequence):
    '''
    Calculate the frequency of each codon in the given DNA sequence.
    :param sequence: DNA sequence.
    :return: Dictionary with codons as keys and their frequencies as
    ↪  values.
    '''
    codons = [sequence[i:i+3] for i in range(0, len(sequence), 3)]
    freq = Counter(codons)
    total_codons = len(codons)

    return {codon: count/total_codons for codon, count in
    ↪  freq.items()}

def optimize_codon_usage(codon_freq, target_host_codon_bias):
    '''
    Optimize codon usage based on the target host's codon bias.
    :param codon_freq: Dictionary of original codon frequencies.
    :param target_host_codon_bias: Dictionary with the target host's
    ↪  preferred codon frequencies.
    :return: Dictionary with optimized codon frequencies.
    '''
```

```python
        optimized_codon_usage = {}
        for codon in codon_freq:
            if codon in target_host_codon_bias:
                optimized_codon_usage[codon] =
                ↪   target_host_codon_bias[codon]
            else:
                optimized_codon_usage[codon] = codon_freq[codon]

        return optimized_codon_usage

def simulate_protein_expression(sequence, codon_usage):
    '''
    Simulate protein expression levels based on optimized codon
    ↪   usage.
    :param sequence: Original DNA sequence.
    :param codon_usage: Optimized codon usage dictionary.
    :return: Estimated expression level (arbitrary units).
    '''
    expression_level = 0
    codons = [sequence[i:i+3] for i in range(0, len(sequence), 3)]

    for codon in codons:
        if codon in codon_usage:
            expression_level += codon_usage[codon]

    return expression_level

# Example sequence and host data
example_sequence = "ATGGCCATTGTAATGGGCCGCTGAAAGGGTGCCCGATAG"
target_host_bias = {
    'ATG': 0.21, 'GCC': 0.15, 'ATT': 0.18, 'GTA': 0.12,
    'ATG': 0.21, 'GGC': 0.25, 'CGC': 0.15, 'TGA': 0.10
}

# Calculate initial codon frequencies
initial_codon_freq = codon_frequency(example_sequence)

# Optimize codon usage
optimized_codons = optimize_codon_usage(initial_codon_freq,
↪   target_host_bias)

# Simulate protein expression level
expression_level = simulate_protein_expression(example_sequence,
↪   optimized_codons)

print("Initial Codon Frequency:", initial_codon_freq)
print("Optimized Codon Usage:", optimized_codons)
print("Simulated Expression Level:", expression_level)
```

This code defines several key functions necessary for the analysis and manipulation of codon usage to enhance protein expression:

- `codon_frequency` computes the frequency of each codon in a DNA sequence, aiding in understanding current usage.

- `optimize_codon_usage` adjusts codon usage based on a target host's codon bias, promoting efficient translation.

- `simulate_protein_expression` estimates protein expression levels using the optimized codon usage profile.

The final block of code demonstrates the calculation of initial codon frequencies, optimization of these frequencies according to host preferences, and estimation of the expected protein expression level using a sample DNA sequence and arbitrary data for a target host.

Chapter 66

Error Correction Algorithms in DNA Synthesis

Below is a Python code snippet that showcases an algorithm for detecting and correcting errors in synthesized DNA sequences. This implementation utilizes a simple consensus sequence approach to identify discrepancies and rectify errors in DNA sequencing data.

```python
def read_sequences(file_path):
    '''
    Reads sequences from a file and stores them in a list.
    :param file_path: Path to the input file containing DNA
     ↪ sequences.
    :return: List of DNA sequences.
    '''
    with open(file_path, 'r') as file:
        sequences = file.readlines()
    # Strip newline characters
    return [seq.strip() for seq in sequences]


def build_consensus(sequences):
    '''
    Constructs a consensus sequence from multiple DNA sequences.
    :param sequences: List of DNA sequences.
    :return: Consensus DNA sequence.
    '''
    if not sequences:
        return ""
```

```python
    consensus = []
    sequence_length = len(sequences[0])

    for i in range(sequence_length):
        # Dictionary to keep count of each nucleotide at position i
        nucleotide_count = {}
        for seq in sequences:
            nucleotide = seq[i]
            if nucleotide in nucleotide_count:
                nucleotide_count[nucleotide] += 1
            else:
                nucleotide_count[nucleotide] = 1

        # Determine the most common nucleotide
        most_common_nucleotide = max(nucleotide_count,
            key=nucleotide_count.get)
        consensus.append(most_common_nucleotide)

    return ''.join(consensus)


def error_correction(sequences, consensus):
    '''
    Corrects errors in sequences based on a consensus sequence.
    :param sequences: List of original DNA sequences.
    :param consensus: Consensus DNA sequence.
    :return: List of corrected DNA sequences.
    '''
    corrected_sequences = []

    for seq in sequences:
        corrected_seq = list(seq)
        for i in range(len(seq)):
            if seq[i] != consensus[i]:
                corrected_seq[i] = consensus[i]
        corrected_sequences.append(''.join(corrected_seq))

    return corrected_sequences


# Example Usage:
input_file_path = 'dna_sequences.txt'
sequences = read_sequences(input_file_path)
consensus_sequence = build_consensus(sequences)
corrected_sequences = error_correction(sequences,
    consensus_sequence)

# Output of corrected sequences
for seq in corrected_sequences:
    print(seq)
```

This code provides a basic implementation of error correction in synthesized DNA sequences through:

- `read_sequences` function reads DNA sequences from a specified file path and returns them as a list.

- `build_consensus` constructs a consensus sequence by iterating through each position in the DNA sequences and determining the most frequent nucleotide.

- `error_correction` compares each sequence with the consensus sequence to detect and correct mismatches.

The example usage at the end demonstrates reading sequences from a file, building a consensus sequence, correcting errors, and printing the corrected sequences.

Chapter 67

High-Throughput Sequencing Error Correction

Below is a Python code snippet that implements error correction algorithms in next-generation sequencing data, focusing on methods like k-mer counting and leveraging error profiles to correct erroneous reads in sequencing datasets.

```
import numpy as np
from collections import defaultdict

def count_kmers(sequence, k):
    '''
    Count k-mers in a sequence.
    :param sequence: Input DNA sequence.
    :param k: Length of k-mer.
    :return: Dictionary with k-mer counts.
    '''
    kmers = defaultdict(int)
    for i in range(len(sequence) - k + 1):
        kmer = sequence[i:i + k]
        kmers[kmer] += 1
    return kmers

def find_error_kmers(kmer_counts, threshold):
    '''
    Identify potential error k-mers.
    :param kmer_counts: Dictionary of k-mer counts.
    :param threshold: Count threshold to identify errors.
    :return: Set of error-prone k-mers.
```

```python
    '''
    error_kmers = {kmer for kmer, count in kmer_counts.items() if
      count < threshold}
    return error_kmers

def correct_errors(reads, error_kmers, k):
    '''
    Correct sequencing errors in reads.
    :param reads: List of DNA reads.
    :param error_kmers: Set of identified error-prone k-mers.
    :param k: Length of k-mer.
    :return: List of corrected reads.
    '''
    corrected_reads = []
    for read in reads:
        new_read = list(read)
        for i in range(len(read) - k + 1):
            kmer = read[i:i + k]
            if kmer in error_kmers:
                # Replace the least frequent k-mer by its most
                  possible correction
                for candidate in generate_kmer_candidates(kmer):
                    if candidate not in error_kmers:
                        new_read[i:i+k] = candidate
                        break
        corrected_reads.append(''.join(new_read))
    return corrected_reads

def generate_kmer_candidates(kmer):
    '''
    Generate possible candidate corrections for a k-mer.
    :param kmer: The erroneous k-mer.
    :return: Generator of candidate k-mers.
    '''
    bases = ['A', 'T', 'C', 'G']
    for i in range(len(kmer)):
        for base in bases:
            if base != kmer[i]:
                yield kmer[:i] + base + kmer[i+1:]

# Example usage with mock data
reads = ["ACGTGGCTA", "ACGTGGTCA", "ACGTGGTTA"]
k = 3
kmer_counts = count_kmers(''.join(reads), k)
error_kmers = find_error_kmers(kmer_counts, threshold=2)
corrected_reads = correct_errors(reads, error_kmers, k)

print("Original Reads:", reads)
print("Corrected Reads:", corrected_reads)
```

This code provides a basic framework for error correction in sequencing data using k-mers:

- `count_kmers` function scans through each read to tally up the frequencies of k-mers.

- `find_error_kmers` identifies k-mers appearing below a defined occurrence threshold, suspecting them as errors.

- `correct_errors` attempts to make corrections by replacing problematic k-mers with the most likely alternatives.

- `generate_kmer_candidates` generates possible single-nucleotide variants of a k-mer to aid in error correction.

The process ultimately adjusts sequences by iterating over k-mers and substituting errors with plausible corrections, enhancing sequence accuracy.

Chapter 68

Data Compression Techniques for Genomics

Below is a Python code snippet that encompasses the core computational elements of genomic data compression, utilizing a differential compression approach to achieve efficient storage without significant loss.

```
import numpy as np
import zlib

def delta_compression(data):
    '''
    Applies delta compression to input genomic data.
    :param data: A list or array of integer genomic data points.
    :return: Delta-compressed data.
    '''
    if len(data) < 2:
        return data
    compressed = np.diff(data, prepend=data[0])
    return compressed

def delta_decompression(compressed_data):
    '''
    Decompresses delta-compressed genomic data.
    :param compressed_data: Delta-compressed genetic data.
    :return: Original decompressed genomic data.
    '''
    decompressed = np.cumsum(compressed_data)
    return decompressed
```

```python
def zlib_compression(data):
    '''
    Compresses data using zlib.
    :param data: The data as bytes or bytearray to be compressed.
    :return: Compressed byte data.
    '''
    return zlib.compress(data)

def zlib_decompression(compressed_data):
    '''
    Decompresses zlib-compressed data.
    :param compressed_data: Compressed byte data.
    :return: Decompressed original byte data.
    '''
    return zlib.decompress(compressed_data)

# Example usage of the compression and decompression functions
genomic_data = np.random.randint(0, 256, size=1000)

# Apply delta compression
delta_compressed_data = delta_compression(genomic_data)
print("Delta Compressed Data:", delta_compressed_data)

# Apply zlib to further compress the delta-compressed data
zlib_compressed_data = 
    zlib_compression(delta_compressed_data.tobytes())
print("Zlib Compressed Data Length:", len(zlib_compressed_data))

# Decompress the zlib data
decompressed_zlib_data = zlib_decompression(zlib_compressed_data)
# Recover the original delta-compressed data
recovered_delta_compressed_data = 
    np.frombuffer(decompressed_zlib_data, dtype=int)

# Decompress the delta data back to original
decompressed_data = 
    delta_decompression(recovered_delta_compressed_data)
print("Decompressed Genomic Data:", decompressed_data)

# Validate compression accuracy
assert np.array_equal(genomic_data, decompressed_data)
```

This code defines several key functions necessary for compressing and decompressing large-scale genomic data:

- `delta_compression` calculates the differences between sequential data points, resulting in a delta-compressed form.
- `delta_decompression` reconstructs the original data by cumulative summing the deltas.

- `zlib_compression` compresses data using the efficient `zlib` library, further reducing the data size.

- `zlib_decompression` restores the data to its delta-compressed state by reversing the `zlib` compression.

The final block of code demonstrates these functions using random genomic data, ensuring that the decompressed data matches the original input data by utilizing assertions for accuracy.

Chapter 69

Secure Sharing of Genomic Data

Below is a Python code snippet that encompasses the core computational elements to ensure privacy and security in the sharing of sensitive genomic information using homomorphic encryption techniques.

```python
from phe import paillier
import numpy as np

def encrypt_genomic_data(public_key, data):
    '''
    Encrypts genomic data using a homomorphic encryption scheme.
    :param public_key: Paillier public key for encryption.
    :param data: List of genomic information to encrypt.
    :return: List of encrypted data.
    '''
    encrypted_data = [public_key.encrypt(item) for item in data]
    return encrypted_data

def decrypt_genomic_data(private_key, encrypted_data):
    '''
    Decrypts genomic data using a homomorphic encryption scheme.
    :param private_key: Paillier private key for decryption.
    :param encrypted_data: List of encrypted genomic information to
    ↪ decrypt.
    :return: List of decrypted data.
    '''
    decrypted_data = [private_key.decrypt(item) for item in
    ↪ encrypted_data]
    return decrypted_data
```

```
def add_encrypted_data(encrypted_data1, encrypted_data2):
    '''
    Adds two lists of encrypted genomic data element-wise.
    :param encrypted_data1: First list of encrypted data.
    :param encrypted_data2: Second list of encrypted data.
    :return: List of resulting encrypted data.
    '''
    return [item1 + item2 for item1, item2 in zip(encrypted_data1,
    ↪  encrypted_data2)]

# Generate public and private keys
public_key, private_key = paillier.generate_paillier_keypair()

# Example genomic data represented as numeric arrays
genomic_data1 = np.array([1.23, 3.45, 5.67, 7.89])
genomic_data2 = np.array([0.12, 0.34, 0.56, 0.78])

# Encrypting the genomic data
encrypted_data1 = encrypt_genomic_data(public_key, genomic_data1)
encrypted_data2 = encrypt_genomic_data(public_key, genomic_data2)

# Adding encrypted data
encrypted_sum = add_encrypted_data(encrypted_data1, encrypted_data2)

# Decrypting the result
decrypted_sum = decrypt_genomic_data(private_key, encrypted_sum)

# Outputs for demonstration
print("Original Genomic Data 1:", genomic_data1)
print("Original Genomic Data 2:", genomic_data2)
print("Decrypted Sum of Genomic Data:", decrypted_sum)
```

This code defines several key functions necessary for encrypting, decrypting, and processing genomic data while ensuring privacy through homomorphic encryption:

- `encrypt_genomic_data` function encrypts genomic information using a given public key to keep data confidential.

- `decrypt_genomic_data` function decrypts the encrypted data using the corresponding private key.

- `add_encrypted_data` demonstrates homomorphic addition on encrypted data, evidencing that computations can be directly performed without decryption.

The final block of code shows the encryption of two genomic datasets, their homomorphic addition, and the decryption of the sum, ensuring the process is both secure and functional.

Chapter 70

Integration of Multi-Omics Data

Below is a Python code snippet that represents a computational framework for integrating genomics, transcriptomics, proteomics, and metabolomics data. The code outlines the core procedures and algorithms necessary for data integration via multi-omics analysis.

```
import pandas as pd
import numpy as np
from sklearn.decomposition import PCA
from sklearn.preprocessing import StandardScaler
from scipy.stats import pearsonr

def load_omics_data(file_paths):
    '''
    Load multi-omics data from specified file paths.
    :param file_paths: Dictionary containing paths to genomics,
     ↪ transcriptomics,
                   proteomics, and metabolomics data files.
    :return: Dictionary of pandas DataFrames for each omics type.
    '''
    data = {}
    for omic_type, path in file_paths.items():
        data[omic_type] = pd.read_csv(path)
    return data

def preprocess_omics_data(data_frames):
    '''
    Preprocess multi-omics data by filling missing values and
     ↪ scaling.
    :param data_frames: Dictionary of DataFrames for each omics
     ↪ type.
```

```python
    :return: Dictionary of preprocessed DataFrames.
    '''
    for omic_type, df in data_frames.items():
        df.fillna(df.mean(), inplace=True)
        scaler = StandardScaler()
        data_frames[omic_type] =
        ↪ pd.DataFrame(scaler.fit_transform(df),
        ↪ columns=df.columns)
    return data_frames

def integrate_omics_data(data_frames):
    '''
    Integrate multi-omics data sets using Principal Component
    ↪ Analysis (PCA).
    :param data_frames: Dictionary of preprocessed DataFrames.
    :return: Integrated DataFrame of principal components across all
    ↪ omics.
    '''
    combined_df = pd.concat(data_frames.values(), axis=1)
    pca = PCA(n_components=3)
    principal_components = pca.fit_transform(combined_df)
    return pd.DataFrame(principal_components, columns=['PC1', 'PC2',
    ↪ 'PC3'])

def correlate_omics_data(integrated_data, omics_types):
    '''
    Calculate correlations between different omic layers in the
    ↪ integrated data.
    :param integrated_data: DataFrame of principal components.
    :param omics_types: List of omics types included in the
    ↪ integration.
    :return: Dictionary of correlation matrices between omics types.
    '''
    correlations = {}
    for i, omic_x in enumerate(omics_types):
        for j, omic_y in enumerate(omics_types[i+1:], i+1):
            corr_matrix = integrated_data.corr(method=pearsonr)
            correlations[f'{omic_x}-{omic_y}'] = corr_matrix
    return correlations

# Example usage
file_paths = {
    'genomics': 'path/to/genomics.csv',
    'transcriptomics': 'path/to/transcriptomics.csv',
    'proteomics': 'path/to/proteomics.csv',
    'metabolomics': 'path/to/metabolomics.csv'
}

omics_data = load_omics_data(file_paths)
preprocessed_data = preprocess_omics_data(omics_data)
integrated_data = integrate_omics_data(preprocessed_data)
correlations = correlate_omics_data(integrated_data,
↪ list(file_paths.keys()))
```

```
print("Integrated Data:\n", integrated_data.head())
print("Correlations:\n", correlations)
```

This code encompasses several core computational functions crucial for the integration of multi-omics data:

- `load_omics_data` function reads and loads omics data from a specified set of file paths.

- `preprocess_omics_data` handles missing values and standardizes omics datasets using mean imputation and scaling.

- `integrate_omics_data` applies Principal Component Analysis (PCA) to reduce dimensionality and integrate data across the omics.

- `correlate_omics_data` computes correlation matrices to explore relationships between integrated omics datasets.

These procedures are foundational for multi-omics analysis, allowing researchers to explore correlations and insights across different biological data types.

Chapter 71

Statistical Methods in Bioinformatics

Below is a Python code snippet that demonstrates how advanced statistical algorithms can be applied to the analysis of genetic data. This example includes functions for performing statistical tests, data normalization, and generating summary statistics.

```python
import numpy as np
import pandas as pd
from scipy import stats
from sklearn.preprocessing import StandardScaler

def load_genetic_data(file_path):
    '''
    Loads genetic data from a CSV file into a pandas DataFrame.
    :param file_path: Path to the CSV file containing genetic data.
    :return: DataFrame with genetic data.
    '''
    return pd.read_csv(file_path)

def normalize_data(data):
    '''
    Normalizes the genetic data using z-score normalization.
    :param data: DataFrame containing the genetic data.
    :return: Normalized genetic data.
    '''
    scaler = StandardScaler()
    normalized_data = scaler.fit_transform(data)
    return pd.DataFrame(normalized_data, columns=data.columns)

def t_test(data, column_a, column_b):
    '''
```

```
    Performs a t-test between two columns of genetic data.
    :param data: DataFrame containing the genetic data.
    :param column_a: First column for the t-test.
    :param column_b: Second column for the t-test.
    :return: t-statistic and p-value.
    '''
    t_stat, p_value = stats.ttest_ind(data[column_a],
    ↪    data[column_b], equal_var=False)
    return t_stat, p_value

def summary_statistics(data, column):
    '''
    Generates summary statistics for a given column in the data.
    :param data: DataFrame containing the genetic data.
    :param column: Column for which to generate summary statistics.
    :return: Dictionary of summary statistics.
    '''
    mean = np.mean(data[column])
    median = np.median(data[column])
    std_dev = np.std(data[column])
    variance = np.var(data[column])

    return {
        'mean': mean,
        'median': median,
        'standard deviation': std_dev,
        'variance': variance
    }

# Example usage: Assume we have a CSV file 'genetic_data.csv' with
↪    columns 'gene_1' and 'gene_2'
file_path = 'genetic_data.csv'    # Replace with your file path
genetic_data = load_genetic_data(file_path)
normalized_genomic_data = normalize_data(genetic_data)

t_stat, p_value = t_test(normalized_genomic_data, 'gene_1',
↪    'gene_2')
summary_stats_gene_1 = summary_statistics(normalized_genomic_data,
↪    'gene_1')

print("T-test between gene_1 and gene_2: t-statistic =", t_stat, ",
↪    p-value =", p_value)
print("Summary Statistics for gene_1:", summary_stats_gene_1)
```

This code snippet provides essential functionalities for analyzing genetic data using statistical methods:

- `load_genetic_data` function reads genetic data from a CSV file into a DataFrame for further analysis.

- `normalize_data` function applies z-score normalization to the genetic dataset, ensuring all features have comparable scales.

- **t_test** performs a statistical comparison between two samples, 'column_a' and 'column_b', using the independent t-test.

- **summary_statistics** function computes basic statistical metrics such as mean, median, standard deviation, and variance for a specified data column.

This example illustrates how Python can be utilized to manage and statistically analyze genetic data efficiently.

Chapter 72

Bayesian Networks in Genetic Analysis

Below is a Python code snippet that demonstrates how Bayesian networks can be applied to model complex genetic relationships. This code involves setting up a Bayesian network for a simple hypothetical genetic system, estimating probabilities, and inferring the likelihood of certain genetic states using the model.

```
import numpy as np
import pandas as pd
from pgmpy.estimators import ParameterEstimator,
    MaximumLikelihoodEstimator
from pgmpy.models import BayesianModel
from pgmpy.inference import VariableElimination
from pgmpy.estimators import BayesianEstimator

# Sample data for constructing the Bayesian network
data = pd.DataFrame(data={
    'GeneA': ['present', 'absent', 'present', 'present', 'absent',
        'absent'],
    'GeneB': ['present', 'present', 'absent', 'present', 'absent',
        'present'],
    'Trait': ['high', 'low', 'high', 'medium', 'low', 'medium']
})

# Define the structure of the Bayesian network
model = BayesianModel([
    ('GeneA', 'Trait'),
    ('GeneB', 'Trait')
])
```

```python
# Fit the model using Maximum Likelihood Estimator
model.fit(data, estimator=MaximumLikelihoodEstimator)

# Perform inference
inference = VariableElimination(model)

# Query to infer the probability of each Trait level given GeneA is
↪   present
prob_trait_given_geneA = inference.query(variables=['Trait'],
↪   evidence={'GeneA': 'present'})

# Output the results
print(prob_trait_given_geneA)

# Query to infer the probability of Trait given both genes are
↪   present
prob_trait_given_genes = inference.query(variables=['Trait'],
↪   evidence={'GeneA': 'present', 'GeneB': 'present'})

# Output the results
print(prob_trait_given_genes)

# Using Bayesian Estimation
model.fit(data, estimator=BayesianEstimator, prior_type="BDeu")
prob_trait_given_genes_bayesian =
↪   inference.query(variables=['Trait'], evidence={'GeneA':
↪   'present', 'GeneB': 'present'})

# Output the Bayesian Estimation results
print(prob_trait_given_genes_bayesian)
```

This code defines the core steps necessary for applying Bayesian networks to model genetic relationships:

- **BayesianModel** sets up the structure of the Bayesian network, defining the dependencies between genes and traits.

- **fit** method uses the data to estimate the conditional probability distributions with both Maximum Likelihood and Bayesian estimation methods.

- **VariableElimination** is used for performing inference tasks to query the likelihood of genetic traits given certain gene expressions.

- The code provides examples of using the network to infer the probabilities of a trait being 'high', 'medium', or 'low' given the presence of certain genes.

The final outputs demonstrate the inferential capability of the Bayesian network model constructed using sample genetic data.

Chapter 73

Markov Chain Monte Carlo Methods in Genetics

Below is a Python code snippet that demonstrates the use of Markov Chain Monte Carlo (MCMC) algorithms for sampling and inference in genetic models, focusing on the implementation of a basic Metropolis-Hastings algorithm.

```python
import numpy as np

def target_distribution(x):
    """
    Define the target distribution that we wish to sample from.
    For demonstration, let's assume a simple Gaussian distribution.
    :param x: The point at which to evaluate the distribution.
    :return: The probability density of the target distribution at
    ↪ x.
    """
    mean = 0
    std_dev = 1
    return np.exp(-0.5 * ((x - mean) / std_dev) ** 2) / (std_dev *
    ↪ np.sqrt(2 * np.pi))

def proposal_distribution(x, step_size=0.5):
    """
    Define the proposal distribution to explore the state space.
    Here, a symmetric Gaussian proposal is used.
    :param x: The current state.
    :param step_size: Standard deviation of the Gaussian proposal
    ↪ distribution.
```

```
    :return: A new proposed state.
    """
    return np.random.normal(x, step_size)

def metropolis_hastings(iterations, initial_value):
    """
    Execute the Metropolis-Hastings algorithm for a given number of
    ↪ iterations.
    :param iterations: Number of samples to generate.
    :param initial_value: Starting point of the Markov chain.
    :return: Samples drawn from the target distribution.
    """
    samples = [initial_value]
    current_x = initial_value

    for _ in range(iterations):
        proposed_x = proposal_distribution(current_x)
        acceptance_ratio = (target_distribution(proposed_x) /
                            target_distribution(current_x))

        if acceptance_ratio >= np.random.rand():
            samples.append(proposed_x)
            current_x = proposed_x
        else:
            samples.append(current_x)

    return samples

# Define parameters for the algorithm
iterations = 10000
initial_value = 0

# Run Metropolis-Hastings to draw samples
samples = metropolis_hastings(iterations, initial_value)

# Analysis of results: simple statistic
mean_sampled = np.mean(samples)
std_dev_sampled = np.std(samples)

print("Sampled Mean: ", mean_sampled)
print("Sampled Standard Deviation: ", std_dev_sampled)
```

This code demonstrates the basic components of the Metropolis-Hastings algorithm used for MCMC sampling:

- `target_distribution` defines the target distribution, which in this example is a Gaussian distribution.

- `proposal_distribution` generates new proposed states using a symmetric Gaussian distribution centered around the current state.

- `metropolis_hastings` combines these elements to perform MCMC sampling, using an acceptance ratio to decide whether to accept a proposed state.

- The main execution block runs the sampling process and provides a simple statistical analysis of the results.

The algorithm efficiently samples from the target distribution, providing an estimate of its mean and standard deviation, which can be used for inference in genetic models.

Chapter 74

High-Performance Computing for Genomics

Below is a Python code snippet that encompasses the concepts of optimizing algorithms for parallel and distributed computing environments. This example demonstrates the use of Python's 'concurrent.futures' module for parallelizing tasks across multiple processors:

```
import concurrent.futures
import numpy as np

def task_function(data_chunk):
    '''
    A sample computational task that performs heavy computation.
    :param data_chunk: A chunk of data to be processed.
    :return: The computed result.
    '''
    # Example heavy computation using numpy
    result = np.sum(np.square(data_chunk))
    return result

def divide_data(data, num_chunks):
    '''
    Split the data into chunks for parallel processing.
    :param data: The data set to divide.
    :param num_chunks: Number of chunks to create.
    :return: A list of data chunks.
    '''
```

```python
    return np.array_split(data, num_chunks)

def parallel_processing(data, num_workers=4):
    '''
    Use parallel processing to perform tasks on data chunks.
    :param data: The complete data set.
    :param num_workers: Number of parallel workers.
    :return: The aggregated result from parallel computation.
    '''
    # Divide data into chunks
    data_chunks = divide_data(data, num_workers)

    # Use ThreadPoolExecutor for parallel processing
    with concurrent.futures.ThreadPoolExecutor(max_workers=num_workers) as executor:
        # Map tasks to workers
        futures = [executor.submit(task_function, chunk) for chunk in data_chunks]

        # Gather results
        results = [future.result() for future in concurrent.futures.as_completed(futures)]

    # Aggregate results
    total_result = np.sum(results)
    return total_result

# Example usage
data_length = 1000000
data = np.random.rand(data_length)
num_workers = 8  # Number of parallel workers

total_result = parallel_processing(data, num_workers)
print("Total Result from Parallel Processing:", total_result)
```

This code snippet includes several key components for optimizing tasks in parallel environments:

- The `task_function` performs a sample computation, intended to mimic a resource-intensive process.

- The `divide_data` function splits a large dataset into smaller chunks that can be processed in parallel.

- The `parallel_processing` function orchestrates the parallel execution of tasks using the 'ThreadPoolExecutor' from the 'concurrent.futures' module.

- This approach takes advantage of multiple CPU cores, improving the efficiency and speed of computationally intensive operations.

The final part of the code demonstrates the implementation with a randomly generated dataset and displays the aggregated result.

Chapter 75

Cloud Computing Architectures for Bioinformatics

Below is a Python code snippet that demonstrates the design and deployment of a scalable cloud computing architecture for handling large-scale genomic data analysis, including managing data ingestion, parallel processing, and result aggregation.

```python
import boto3
from concurrent.futures import ThreadPoolExecutor
import numpy as np

def ingest_genomic_data(s3_bucket, data_key):
    '''
    Download genomic data from an S3 bucket.
    :param s3_bucket: Name of the S3 bucket.
    :param data_key: Key for the data file in S3.
    :return: Genomic data as a numpy array.
    '''
    s3 = boto3.client('s3')
    response = s3.get_object(Bucket=s3_bucket, Key=data_key)
    data = np.load(response['Body'])
    return data

def genomic_analysis(data_chunk):
    '''
    Analyze a chunk of genomic data.
    :param data_chunk: Part of genomic dataset.
    :return: Analysis result.
    '''
```

```python
    # Placeholder for genomic analysis
    result = np.mean(data_chunk)  # Simplified example
    return result

def parallel_processing(data, num_threads):
    '''
    Process genomic data in parallel using threads.
    :param data: Complete genomic data.
    :param num_threads: Number of threads to use in processing.
    :return: List of results from each thread.
    '''
    chunk_size = len(data) // num_threads
    chunks = [data[i:i + chunk_size] for i in range(0, len(data),
        chunk_size)]

    with ThreadPoolExecutor(max_workers=num_threads) as executor:
        results = list(executor.map(genomic_analysis, chunks))

    return results

def aggregate_results(results):
    '''
    Aggregate results from parallel processing.
    :param results: List of results from each thread.
    :return: Aggregated result.
    '''
    aggregated_result = np.sum(results)
    return aggregated_result

# Example of orchestrating the workflow
def main(s3_bucket, data_key):
    '''
    Main orchestration function for genomic data analysis workflow.
    :param s3_bucket: Name of the S3 bucket containing data.
    :param data_key: Data identifier in S3.
    :return: Final aggregated result.
    '''
    data = ingest_genomic_data(s3_bucket, data_key)
    results = parallel_processing(data, num_threads=4)
    final_result = aggregate_results(results)

    return final_result

# Workflow execution
s3_bucket = 'genomic-data-bucket'
data_key = 'large-genomic-dataset.npy'
final_result = main(s3_bucket, data_key)
print("Final Aggregated Result:", final_result)
```

This code outlines a complete workflow for scalable genomic data analysis using cloud infrastructure:

- `ingest_genomic_data` function handles the retrieval of genomic data from a cloud storage (Amazon S3 in this case).

- `genomic_analysis` serves as a proxy for more complex analyses, simplified here as a mean calculation.

- `parallel_processing` utilizes Python's `ThreadPoolExecutor` to distribute the workload across multiple threads, illustrating parallel processing capabilities.

- `aggregate_results` compiles individual thread results into a comprehensive analysis output.

- `main` orchestrates the entire workflow from data ingestion to final computation, reflecting the core of scalable computing solutions.

The script assumes an AWS environment, using Boto3 for S3 interactions, demonstrating scalability and flexibility suitable for large-scale genomic data analysis projects.

Chapter 76

GPU Acceleration of Bioinformatics Algorithms

Below is a Python code snippet that demonstrates leveraging GPU computing to accelerate a computationally intensive genetic algorithm. The code showcases key steps from configuring GPU resources to executing the algorithm efficiently.

```python
import numpy as np
import cupy as cp  # CuPy serves as a GPU-accelerated drop-in
                   # replacement for NumPy

def initialize_population(size, dimensions):
    '''
    Initialize a population for the genetic algorithm.
    :param size: Number of individuals in the population.
    :param dimensions: Number of dimensions each individual has.
    :return: Initial population.
    '''
    return cp.random.rand(size, dimensions)

def evaluate_fitness(population):
    '''
    Example fitness evaluation function.
    :param population: Current population array.
    :return: Fitness scores.
    '''
    # Here, we're simply summing the individual's dimensions to form
    # a score
    return cp.sum(population, axis=1)
```

```python
def select_mating_pool(population, fitness, num_parents):
    '''
    Select a mating pool based on fitness.
    :param population: Current population array.
    :param fitness: Array of fitness scores.
    :param num_parents: Number of parents to select.
    :return: Array of selected parents.
    '''
    parents = cp.empty((num_parents, population.shape[1]))
    for parent_idx in range(num_parents):
        max_fitness_idx = cp.argmax(fitness)
        parents[parent_idx, :] = population[max_fitness_idx, :]
        fitness[max_fitness_idx] = -cp.inf  # exclude this
        ↪    individual from being selected again
    return parents

def crossover(parents, offspring_size):
    '''
    Perform crossover operation.
    :param parents: Array of selected parents.
    :param offspring_size: Number of offspring to generate.
    :return: Array of offsprings.
    '''
    offspring = cp.empty(offspring_size)
    crossover_point = cp.uint32(offspring_size[1] / 2)

    for k in range(offspring_size[0]):
        parent1_idx = k % parents.shape[0]
        parent2_idx = (k + 1) % parents.shape[0]
        offspring[k, 0:crossover_point] = parents[parent1_idx,
        ↪    0:crossover_point]
        offspring[k, crossover_point:] = parents[parent2_idx,
        ↪    crossover_point:]

    return offspring

def mutation(offspring_crossover, mutation_rate):
    '''
    Apply mutation to the offspring.
    :param offspring_crossover: Array of offspring from crossover.
    :param mutation_rate: Mutation rate determining the extent of
    ↪    mutations.
    :return: Mutated offspring.
    '''
    for idx in range(offspring_crossover.shape[0]):
        for gene in range(offspring_crossover.shape[1]):
            random_value = cp.random.rand()
            if random_value < mutation_rate:
                random_value_addition = cp.random.uniform(-1.0, 1.0)
                offspring_crossover[idx, gene] +=
                ↪    random_value_addition
    return offspring_crossover
```

```
# Parameters
population_size = 100
num_generations = 100
dimensions = 10
num_parents = 50
mutation_rate = 0.1

# Initialize the population
population = initialize_population(population_size, dimensions)

# Begin the genetic algorithm loop
for generation in range(num_generations):
    fitness = evaluate_fitness(population)
    parents = select_mating_pool(population, fitness, num_parents)
    offspring_crossover = crossover(parents, (population_size -
    ↪    parents.shape[0], dimensions))
    offspring_mutation = mutation(offspring_crossover,
    ↪    mutation_rate)
    population[0:parents.shape[0], :] = parents
    population[parents.shape[0]:, :] = offspring_mutation

# Best solution
final_fitness = evaluate_fitness(population)
best_match_idx = cp.argmax(final_fitness)
best_solution = population[best_match_idx, :]
print("Best solution:", cp.asnumpy(best_solution))
```

This code snippet implements a GPU-accelerated genetic algorithm, which involves several computational steps represented by distinct functions:

- initialize_population generates the initial population of solutions using random values.

- evaluate_fitness calculates the fitness of each individual in the population based on a simple summation of their values.

- select_mating_pool selects the most fit individuals to become parents for the next generation, utilizing GPU capabilities to manage large data efficiently.

- crossover mixes parental information to create new offspring, promoting genetic diversity.

- mutation adds random alterations to offspring, enhancing variability across generations.

The final output provides the most optimal solution determined after the specified number of generations.

Chapter 77

Reinforcement Learning in Genetic Engineering

Below is a Python code snippet that demonstrates how reinforcement learning can be applied to optimize genetic engineering experiments. This code utilizes the gym environment to simulate experiments, with a reinforcement learning agent that learns to optimize outcomes through trial and error.

```python
import numpy as np
import gym
from gym import spaces

class GeneticEnvironment(gym.Env):
    '''
    A custom OpenAI Gym environment for simulating genetic
    ↪  engineering experiments.
    '''
    def __init__(self, n_genes=10):
        super(GeneticEnvironment, self).__init__()
        self.n_genes = n_genes
        self.action_space = spaces.Discrete(n_genes)  # Actions:
        ↪  which gene to alter
        self.observation_space = spaces.Box(low=0, high=1,
        ↪  shape=(n_genes,), dtype=np.float32)

        # Initial random state of the genes
        self.state = np.random.rand(self.n_genes)
```

```python
def step(self, action):
    '''
    Modifies the state of a selected gene and computes a reward.
    :param action: The index of the gene to alter.
    :return: Tuple containing new state, reward, done flag, and
    ↪  info.
    '''
    # Alter the selected gene (for illustration, add random
    ↪  noise)
    self.state[action] += np.random.normal(0, 0.1)
    self.state = np.clip(self.state, 0, 1)  # Ensure gene state
    ↪  stays within bounds

    # Reward is a function of how optimal the gene states are
    # For demonstration purposes, let's aim for all gene values
    ↪  to be 0.5
    reward = -np.sum((self.state - 0.5) ** 2)

    # Done when reaching a near-optimal solution
    done = reward > -0.1

    return self.state, reward, done, {}

def reset(self):
    '''
    Resets the environment to a new random initial state.
    :return: Initial state.
    '''
    self.state = np.random.rand(self.n_genes)
    return self.state

def render(self, mode='human'):
    '''
    Prints out the current state of the environment.
    '''
    print(f"Current state: {self.state}")

# Simple reinforcement learning agent
class RLAgent:
    def __init__(self, env):
        self.env = env
        self.n_actions = env.action_space.n

    def choose_action(self, state):
        '''
        Chooses an action using a simple strategy. For demonstration
        ↪  purposes, use random choice.
        :param state: The current state of the environment.
        :return: Selected action.
        '''
        return np.random.choice(self.n_actions)

    def train(self, n_episodes=1000):
```

```python
        '''
        Trains the agent over multiple episodes.
        :param n_episodes: Number of episodes to run training for.
        '''
        for episode in range(n_episodes):
            state = self.env.reset()
            done = False
            total_reward = 0

            while not done:
                action = self.choose_action(state)
                next_state, reward, done, _ = self.env.step(action)
                total_reward += reward
                state = next_state

            print(f"Episode {episode + 1}: Total Reward:
            ↪ {total_reward}")

if __name__ == "__main__":
    # Create environment and agent
    env = GeneticEnvironment()
    agent = RLAgent(env)

    # Train agent
    agent.train()
```

This code defines a reinforcement learning setup to optimize genetic engineering experiments:

- The `GeneticEnvironment` class simulates the genetic experiment, with states representing gene expressions and rewards indicating how close the gene expressions are to an optimal state.

- The `step` function modifies a specific gene and calculates a reward based on its new expression.

- The `RLAgent` class is a simple reinforcement learning agent that interacts with the environment, choosing actions and learning from the rewards received.

- The `train` method conducts multiple episodes to improve the agent's performance in optimizing the genetic states.

- Example execution demonstrates how the agent attempts to maximize the reward through iterative episodes.

Chapter 78

Evolutionary Algorithms for Protein Engineering

Below is a Python code snippet that illustrates the use of evolutionary computation to design proteins with desired properties. The example leverages a genetic algorithm to optimize protein sequences based on a hypothetical fitness function evaluating the stability and binding affinity of a protein.

```python
import numpy as np
import random
from typing import List, Tuple

# Protein sequence amino acids (simplified)
amino_acids = 'ACDEFGHIKLMNPQRSTVWY'

def random_protein_sequence(length: int) -> str:
    """Generate a random protein sequence of given length."""
    return ''.join(random.choice(amino_acids) for _ in
        range(length))

def calculate_fitness(sequence: str) -> float:
    """
    Fitness function that evaluates the 'stability' and 'binding
        affinity'
    of the protein sequence. For demonstration, a toy function.
    """
    # Toy example, real scenarios need a biophysical model
    stability_score = sequence.count('A')
```

```python
        binding_affinity_score = sequence.count('W')
        return stability_score + 2 * binding_affinity_score

    def mutate_sequence(sequence: str, mutation_rate: float) -> str:
        """Mutate a protein sequence based on the mutation rate."""
        new_sequence = list(sequence)
        for i in range(len(sequence)):
            if random.random() < mutation_rate:
                new_sequence[i] = random.choice(amino_acids)
        return ''.join(new_sequence)

    def crossover(seq1: str, seq2: str) -> Tuple[str, str]:
        """Generate two offspring sequences by crossover."""
        if len(seq1) != len(seq2):
            raise ValueError("Sequences must have the same length")

        crossover_point = random.randint(1, len(seq1) - 1)
        child1 = seq1[:crossover_point] + seq2[crossover_point:]
        child2 = seq2[:crossover_point] + seq1[crossover_point:]

        return child1, child2

    def genetic_algorithm(population_size: int, sequence_length: int,
                          generations: int, mutation_rate: float) ->
                       ↪    str:
        """
        Main genetic algorithm for evolving protein sequences.

        :param population_size: Number of sequences in the population.
        :param sequence_length: Length of each protein sequence.
        :param generations: Number of generations to run the evolution.
        :param mutation_rate: Probability of mutation at each position.
        :return: The best sequence found.
        """
        # Initialize population with random sequences
        population = [random_protein_sequence(sequence_length) for _ in
                   ↪    range(population_size)]

        for generation in range(generations):
            # Evaluate fitness of each sequence
            fitness_scores = [(sequence, calculate_fitness(sequence))
                           ↪    for sequence in population]

            # Sort by decreasing fitness
            fitness_scores.sort(key=lambda x: x[1], reverse=True)

            # Take top half as parents for the next generation
            selected = fitness_scores[:population_size // 2]

            # Generate new population through crossover and mutation
            new_population = []
            while len(new_population) < population_size:
                parent1 = random.choice(selected)[0]
```

```
            parent2 = random.choice(selected)[0]
            child1, child2 = crossover(parent1, parent2)
            new_population.append(mutate_sequence(child1,
            ↪    mutation_rate))
            new_population.append(mutate_sequence(child2,
            ↪    mutation_rate))

        population = new_population[:population_size]  # Preserve
        ↪    population size

    # Return the best sequence found
    best_sequence = max(population, key=calculate_fitness)
    return best_sequence

# Example parameters
population_size = 100
sequence_length = 10
generations = 50
mutation_rate = 0.05

# Run genetic algorithm
best_protein_sequence = genetic_algorithm(population_size,
↪    sequence_length, generations, mutation_rate)
print("Best protein sequence found:", best_protein_sequence)
```

This code defines and implements the basic components of a genetic algorithm for protein design:

- random_protein_sequence generates random amino acid sequences.

- calculate_fitness provides a toy fitness function for assessing protein properties.

- mutate_sequence introduces random mutations into a sequence.

- crossover performs crossover operations to combine sequences and create diversity in the population.

- genetic_algorithm manages the main genetic algorithm loop, evolving the sequences over a number of generations to optimize the desired properties of stability and binding affinity.

The final section demonstrates executing the genetic algorithm with specified parameters to find an optimized protein sequence.

Chapter 79

Genetic Programming in Synthetic Biology

Below is a Python code snippet that encompasses the core computational elements for evolving programs to design synthetic biological systems using genetic programming.

```python
import random

class GeneticProgramming:
    def __init__(self, functions, terminals, population_size=100,
                 generations=50,
                 mutation_rate=0.1, crossover_rate=0.9):
        self.functions = functions
        self.terminals = terminals
        self.population_size = population_size
        self.generations = generations
        self.mutation_rate = mutation_rate
        self.crossover_rate = crossover_rate
        self.population = self._initialize_population()

    def _initialize_population(self):
        '''
        Initialize a random population of programs.
        '''
        return [self._random_program() for _ in
                range(self.population_size)]

    def _random_program(self, depth=3):
        '''
        Recursively build a random program.
        :param depth: Maximum depth of program tree.
        :return: Random program.
```

```
        '''
        if depth == 0 or (depth > 1 and random.random() < 0.5):
            return random.choice(self.terminals)
        func = random.choice(self.functions)
        return [func] + [self._random_program(depth - 1) for _ in
        ↪    range(func.arity)]

    def evaluate(self, program, inputs):
        '''
        Evaluate a program on given inputs.
        :param program: The program to evaluate.
        :param inputs: Input values.
        :return: Evaluated result.
        '''
        if isinstance(program, str):
            # Terminal case: return input directly
            return inputs.get(program, program)
        func = self.functions[program[0]]
        return func(*[self.evaluate(arg, inputs) for arg in
        ↪    program[1:]])

    def fitness(self, program):
        '''
        Calculate fitness of a program.
        :param program: Program tree to evaluate.
        :return: Fitness score.
        '''
        return -abs(self.evaluate(program, {'x': 5}) - 42)  #
        ↪    Example target function

    def mutate(self, program):
        '''
        Perform mutation on the program.
        :param program: Original program.
        :return: Mutated program.
        '''
        if random.random() < 0.1 or not isinstance(program, list):
            return self._random_program(depth=2)
        else:
            func, *args = program
            return [func] + [self.mutate(arg) for arg in args]

    def crossover(self, parent1, parent2):
        '''
        Perform crossover between two parent programs.
        :param parent1: First parent program.
        :param parent2: Second parent program.
        :return: Combined program.
        '''
        if not isinstance(parent1, list) or not isinstance(parent2,
        ↪    list):
            return parent1 if random.random() < 0.5 else parent2
        func1, *args1 = parent1
```

```
            func2, *args2 = parent2
            if random.random() < 0.5:
                return [func1] + [self.crossover(arg1, arg2) for arg1,
                ↪    arg2 in zip(args1, args2)]
            else:
                return [func2] + args2

    def evolve(self):
        '''
        Evolve the population over generations.
        '''
        for _ in range(self.generations):
            ranked_population = sorted(self.population,
            ↪    key=self.fitness, reverse=True)
            new_population = ranked_population[:self.population_size
            ↪    // 10]  # Select top 10%

            while len(new_population) < self.population_size:
                if random.random() < self.crossover_rate:
                    parent1 = random.choice(ranked_population)
                    parent2 = random.choice(ranked_population)
                    offspring = self.crossover(parent1, parent2)
                else:
                    parent = random.choice(ranked_population)
                    offspring = self.mutate(parent)
                new_population.append(offspring)

            self.population = new_population

# Example usage
functions = {
    'add': lambda x, y: x + y, 'sub': lambda x, y: x - y,
    'mul': lambda x, y: x * y, 'div': lambda x, y: x / y,
}

terminals = ['x', 1, 2, 3, 4]

gp = GeneticProgramming(functions, terminals)
gp.evolve()

# Output the best program
best_program = max(gp.population, key=gp.fitness)
print("Best Program:", best_program)
```

This code outlines the implementation of a basic genetic programming framework:

- The `GeneticProgramming` class manages population initialization, program evaluation, and evolutionary operations.

- The `evaluate` function applies genetic programming operations to compute outcomes based on a provided input set.

- The `fitness` function is a metric evaluating the efficiency of a program by comparing it against a target function.

- A combination of `mutate` and `crossover` functions are used for genetic variations between generations.

The final block of the script demonstrates usage by evolving a population towards a predefined target.

Chapter 80

Optimization Algorithms in Metabolic Engineering

Below is a Python code snippet that encompasses the core computational elements for applying optimization techniques to enhance production yields in metabolic pathways, including the definition of the metabolic network, calculation of production yields, and optimization procedure.

```python
import numpy as np
from scipy.optimize import linprog

def define_metabolic_network(stoichiometry_matrix, reaction_bounds):
    '''
    Define the metabolic network for optimization.
    :param stoichiometry_matrix: Matrix representing reaction
        stoichiometries.
    :param reaction_bounds: Bounds for each reaction in the network.
    :return: Structured network model.
    '''
    return {'S': stoichiometry_matrix, 'bounds': reaction_bounds}

def calculate_production_yield(flux_distribution, product_indices):
    '''
    Calculate the production yield based on the flux distribution.
    :param flux_distribution: Optimized reaction fluxes.
    :param product_indices: Indices of the production reactions.
    :return: Total production yield.
    '''
```

```python
    return np.sum(flux_distribution[product_indices])

def optimize_metabolic_network(metabolic_net, objective_func):
    '''
    Optimize the metabolic network to maximize production yield.
    :param metabolic_net: Network structure containing stoichiometry
    ↪    and bounds.
    :param objective_func: Objective function coefficients for the
    ↪    linear program.
    :return: Optimized flux distribution.
    '''
    result = linprog(
        c=-objective_func,
        A_eq=metabolic_net['S'],
        b_eq=np.zeros(metabolic_net['S'].shape[0]),
        bounds=metabolic_net['bounds'],
        method='highs'
    )
    if result.success:
        return result.x
    else:
        raise ValueError('Optimization failed')

# Example of defining a simple metabolic network
stoichiometry_matrix = np.array([
    [-1,  1,  0,  0],
    [ 0, -1,  1,  0],
    [ 0,  0, -1,  1]
])

reaction_bounds = [
    (0, 10),   # Bounds for reaction 1
    (0, 10),   # Bounds for reaction 2
    (0, 10),   # Bounds for reaction 3
    (0, 10)    # Bounds for reaction 4
]

objective_func = np.array([0, 0, 0, 1])   # Maximizing production
↪    from reaction 4

# Setting up the metabolic network
network = define_metabolic_network(stoichiometry_matrix,
↪    reaction_bounds)

# Performing optimization
optimized_fluxes = optimize_metabolic_network(network,
↪    objective_func)

# Yield calculation
product_indices = [3]   # Assuming reaction 4 is the production
↪    reaction
production_yield = calculate_production_yield(optimized_fluxes,
↪    product_indices)
```

```
print("Optimized Flux Distribution:", optimized_fluxes)
print("Production Yield:", production_yield)
```

This code defines several key functions necessary for the optimization of metabolic pathways:

- `define_metabolic_network` function sets up the metabolic network model using stoichiometry and reaction bounds.

- `calculate_production_yield` computes the yield based on optimized fluxes from the network's output reactions.

- `optimize_metabolic_network` applies linear programming to find flux distributions that maximize the production yield of the target metabolites.

The final block of code provides an example of defining and optimizing a simple metabolic network for maximum production yield with dummy data.

Chapter 81

Swarm Intelligence in Biological Systems

Below is a Python code snippet that exemplifies the use of swarm intelligence algorithms, particularly Ant Colony Optimization (ACO), to model and analyze complex biological behaviors such as resource foraging and pathfinding.

```python
import numpy as np
import random

class Ant:
    def __init__(self, num_nodes):
        self.visited_nodes = []
        self.current_node = random.randint(0, num_nodes - 1)
        self.path_distance = 0
        self.visited_nodes.append(self.current_node)

    def visit_node(self, node, distance):
        self.visited_nodes.append(node)
        self.path_distance += distance
        self.current_node = node

    def path_complete(self, num_nodes):
        return len(self.visited_nodes) == num_nodes

class AntColonyOptimization:
    def __init__(self, graph, num_ants, num_iterations,
                 decay_factor, alpha=1, beta=1):
        """
        :param graph: Distance matrix representing the graph.
        :param num_ants: Number of ants in the colony.
        :param num_iterations: Number of iterations to run.
```

```
    :param decay_factor: Pheromone decay per iteration.
    :param alpha: Influence of pheromone on direction.
    :param beta: Influence of distance on direction.
    """
    self.graph = graph
    self.num_ants = num_ants
    self.num_iterations = num_iterations
    self.decay_factor = decay_factor
    self.alpha = alpha
    self.beta = beta
    self.num_nodes = len(graph)
    self.pheromones = np.ones((self.num_nodes, self.num_nodes))
    ↪    / self.num_nodes

def run(self):
    shortest_path = None
    shortest_distance = float('inf')
    for _ in range(self.num_iterations):
        ants = [Ant(self.num_nodes) for _ in
        ↪    range(self.num_ants)]
        for ant in ants:
            while not ant.path_complete(self.num_nodes):
                next_node = self.choose_next_node(ant)
                ant.visit_node(next_node,
                ↪    self.graph[ant.current_node][next_node])
        self.update_pheromones(ants)

        for ant in ants:
            if ant.path_distance < shortest_distance:
                shortest_path = ant.visited_nodes
                shortest_distance = ant.path_distance
    return shortest_path, shortest_distance

def choose_next_node(self, ant):
    pheromone = np.copy(self.pheromones[ant.current_node])
    probability_distribution =
    ↪    self.get_probability_distribution(ant.current_node,
    ↪    pheromone)
    next_node = self.random_choice(ant.visited_nodes,
    ↪    probability_distribution)
    return next_node

def get_probability_distribution(self, from_node, pheromone):
    total = 0.0
    for to_node in range(self.num_nodes):
        if to_node not in pheromone:
            pheromone[to_node] =
            ↪    (self.pheromones[from_node][to_node] **
            ↪    self.alpha) * \
                                ((1.0 /
                                ↪    self.graph[from_node][to_node])
                                ↪    ** self.beta)
            total += pheromone[to_node]
```

```
        return [pheromone[node] / total for node in
        ↪ range(self.num_nodes)]

    def random_choice(self, visited, probability_distribution):
        cumulative_distribution =
        ↪ np.cumsum(probability_distribution)
        choice = random.random()
        for i, probability in enumerate(cumulative_distribution):
            if choice <= probability and i not in visited:
                return i

    def update_pheromones(self, ants):
        self.pheromones *= (1 - self.decay_factor)
        for ant in ants:
            for i in range(len(ant.visited_nodes) - 1):
                from_node = ant.visited_nodes[i]
                to_node = ant.visited_nodes[i + 1]
                self.pheromones[from_node][to_node] += 1.0 /
                ↪ ant.path_distance

# Test data: A distance matrix representing a simple graph
graph = np.array([
    [0, 2, 2, 3],
    [2, 0, 1, 3],
    [2, 1, 0, 2],
    [3, 3, 2, 0]
])

aco = AntColonyOptimization(graph=graph, num_ants=5,
↪ num_iterations=100, decay_factor=0.5)
shortest_path, shortest_distance = aco.run()

print("Shortest path found:", shortest_path)
print("Shortest path distance:", shortest_distance)
```

This code demonstrates the core implementation of Ant Colony Optimization for solving a pathfinding problem in a graph:

- `Ant` class defines the structure for individual ants including tracking their visited nodes and path distance.

- `AntColonyOptimization` class performs the primary algorithm processes: initializing pheromones, running the ant colony iterations, and updating pheromones.

- `choose_next_node` and `random_choice` methods determine the next node to visit based on pheromones and graph distances.

- `update_pheromones` method adjusts the pheromone trails to reinforce shorter paths.

The example graph is a simple adjacency matrix, and the algorithm tests for the shortest path using specified parameters, demonstrating effective modeling of biological systems using swarm intelligence.

Chapter 82

Bioinformatics Workflow Automation

Below is a Python code snippet that demonstrates the automation of bioinformatics workflows using Python tools and libraries. This example includes setting up and executing a pipeline for processing genomic data, which includes steps like quality control, data transformation, and analysis.

```python
import os
import subprocess
from multiprocessing import Pool

def execute_command(command):
    '''Execute a shell command and return the output.'''
    try:
        output = subprocess.check_output(command, shell=True,
            stderr=subprocess.STDOUT)
        return output.decode()
    except subprocess.CalledProcessError as e:
        return f"An error occurred: {e.output.decode()}"

def quality_control(fastq_file, output_dir):
    '''Run quality control on raw FASTQ files using FastQC.'''
    command = f"fastqc {fastq_file} -o {output_dir}"
    return execute_command(command)

def alignment(fastq_file, reference_genome, output_dir):
    '''Align sequencing reads to a reference genome using BWA.'''
    sam_output = os.path.join(output_dir,
        f"{os.path.basename(fastq_file)}.sam")
    command = f"bwa mem {reference_genome} {fastq_file} > 
        {sam_output}"
```

```python
    return execute_command(command)

def convert_sam_to_bam(sam_file, output_dir):
    '''Convert SAM file to BAM format using Samtools.'''
    bam_output = os.path.join(output_dir,
        f"{os.path.splitext(sam_file)[0]}.bam")
    command = f"samtools view -S -b {sam_file} > {bam_output}"
    return execute_command(command)

def sort_bam(bam_file, output_dir):
    '''Sort BAM file using Samtools.'''
    sorted_bam_output = os.path.join(output_dir,
        f"sorted_{os.path.basename(bam_file)}")
    command = f"samtools sort {bam_file} -o {sorted_bam_output}"
    return execute_command(command)

def variant_calling(sorted_bam_file, reference_genome, output_dir):
    '''Call genetic variants using a variant caller like GATK or
        Freebayes.'''
    vcf_output = os.path.join(output_dir,
        f"{os.path.basename(sorted_bam_file)}.vcf")
    command = f"freebayes -f {reference_genome} {sorted_bam_file} >
        {vcf_output}"
    return execute_command(command)

def pipeline(fastq_file, reference_genome, output_dir):
    '''Run the entire bioinformatics workflow pipeline.'''
    qc_result = quality_control(fastq_file, output_dir)
    print(f"Quality Control Result: {qc_result}")

    alignment_result = alignment(fastq_file, reference_genome,
        output_dir)
    print(f"Alignment Result: {alignment_result}")

    sam_file = os.path.join(output_dir,
        f"{os.path.basename(fastq_file)}.sam")
    bam_conversion_result = convert_sam_to_bam(sam_file, output_dir)
    print(f"SAM to BAM Conversion Result: {bam_conversion_result}")

    bam_file = os.path.join(output_dir,
        f"{os.path.splitext(sam_file)[0]}.bam")
    sorting_result = sort_bam(bam_file, output_dir)
    print(f"BAM Sorting Result: {sorting_result}")

    sorted_bam_file = os.path.join(output_dir,
        f"sorted_{os.path.basename(bam_file)}")
    variant_result = variant_calling(sorted_bam_file,
        reference_genome, output_dir)
    print(f"Variant Calling Result: {variant_result}")

if __name__ == "__main__":
    # Example usage
    input_fastq = "sample.fastq"
```

```
output_directory = "./pipeline_output"
reference = "reference_genome.fa"

if not os.path.exists(output_directory):
    os.makedirs(output_directory)

pipeline(input_fastq, reference, output_directory)
```

This script automates various stages of genomic data processing using command-line tools integrated with Python:

- The `execute_command` function is a utility function to execute shell commands and handle errors.

- The `quality_control` function runs a quality check on the input FASTQ file using *FastQC*.

- `alignment` aligns sequencing reads to a reference genome using *BWA*.

- The `convert_sam_to_bam` function converts SAM files to BAM format using *Samtools*.

- `sort_bam` sorts BAM files, preparing them for variant calling.

- The `variant_calling` function detects variants using a tool like *Freebayes*.

- Finally, the `pipeline` function orchestrates the various stages, applying each in sequence to process genomic data efficiently.

The `pipeline` illustrates how multiple bioinformatics tools can be integrated into a coherent automated workflow, facilitating more efficient data analysis.

Chapter 83

Semantic Web in Genomics

Below is a Python code snippet that demonstrates how semantic web technologies can be applied to integrate and query genomic data using RDF (Resource Description Framework) and SPARQL (SPARQL Protocol and RDF Query Language).

```
from rdflib import Graph, Namespace, Literal, RDF, URIRef
from rdflib.namespace import FOAF, DC

# Define the namespaces
ex = Namespace("http://example.org/")

# Initialize a graph
g = Graph()

# Create genomic data resources
genome = URIRef(ex['genome/Homo_sapiens'])
gene1 = URIRef(ex['gene/BRCA1'])
gene2 = URIRef(ex['gene/TP53'])

# Add data to the graph
g.add((genome, RDF.type, ex.Genome))
g.add((genome, ex.hasGene, gene1))
g.add((genome, ex.hasGene, gene2))

g.add((gene1, RDF.type, ex.Gene))
g.add((gene1, DC.title, Literal("BRCA1 gene")))
g.add((gene1, ex.associatedWith, Literal("Breast cancer
    susceptibility")))

g.add((gene2, RDF.type, ex.Gene))
```

```
g.add((gene2, DC.title, Literal("TP53 gene")))
g.add((gene2, ex.associatedWith, Literal("Tumor protein p53")))

# Define a function to execute SPARQL queries
def query_genomic_data(g):
    '''
    Runs a SPARQL query to extract information about genes related
    ↪ to cancer.
    :param g: The RDF graph containing genomic data.
    :return: Query results.
    '''
    # SPARQL query to retrieve genes associated with cancer
    query = """
    SELECT ?gene ?title
    WHERE {
      ?genome ex:hasGene ?gene .
      ?gene dc:title ?title .
      FILTER regex(?title, "cancer", "i")
    }
    """
    results = g.query(query, initNs={"ex": ex, "dc": DC})
    return results

# Run the query and print results
results = query_genomic_data(g)
for row in results:
    print(f"Gene URI: {row.gene}, Title: {row.title}")
```

This code snippet showcases the integration and querying of genomic data using semantic web technologies, involving:

- Creation of RDF graph with 'rdflib' and defining namespaces for genomic entities.

- Adding genomic data related to BRCA1 and TP53 genes with their associated cancer types.

- Implementation of a function query_genomic_data to perform a SPARQL query that retrieves genes relevant to cancer.

- Execution of the query and printing the results that match the specified conditions.

The example demonstrates how RDF and SPARQL can be utilized to semantically represent and query complex bioinformatics data, facilitating effective data integration and retrieval.

Chapter 84

Ontology Development for Genetic Data

Below is a Python code snippet that encompasses the core computational elements for designing ontologies to standardize and facilitate genetic data sharing. The example utilizes the 'rdflib' library for creating and handling RDF data, which is foundational in constructing ontologies.

```
from rdflib import Graph, Literal, RDF, RDFS, OWL, Namespace, URIRef

def create_ontology():
    '''
    Function to create a simple genetic data sharing ontology using
    ↪ RDFLib.
    :return: An RDF graph representing the ontology.
    '''
    # Define a namespace for the ontology
    GENETICS = Namespace("http://example.org/genetics#")

    # Create a new RDF graph
    graph = Graph()

    # Declare classes in the ontology
    graph.add((GENETICS.Gene, RDF.type, OWL.Class))
    graph.add((GENETICS.Protein, RDF.type, OWL.Class))
    graph.add((GENETICS.Disease, RDF.type, OWL.Class))

    # Declare properties in the ontology
    graph.add((GENETICS.encodes, RDF.type, OWL.ObjectProperty))
    graph.add((GENETICS.hasName, RDF.type, OWL.DatatypeProperty))
    graph.add((GENETICS.isAssociatedWith, RDF.type,
    ↪ OWL.ObjectProperty))
```

```
    # Add labels and relationships
    graph.add((GENETICS.Gene, RDFS.label, Literal("Gene")))
    graph.add((GENETICS.Protein, RDFS.label, Literal("Protein")))
    graph.add((GENETICS.encodes, RDFS.label, Literal("encodes")))

    # Define relationships
    graph.add((GENETICS.Gene, GENETICS.encodes, GENETICS.Protein))
    graph.add((GENETICS.Gene, GENETICS.hasName, Literal("BRCA1")))
    graph.add((GENETICS.Gene, GENETICS.isAssociatedWith,
    ↪    GENETICS.Disease))

    return graph

def serialize_ontology(graph, format='turtle'):
    '''
    Serializes the RDF graph into a specified format.
    :param graph: The RDF graph representing the ontology.
    :param format: The format to serialize to (e.g., 'turtle',
    ↪    'xml').
    :return: A string containing the serialized ontology.
    '''
    return str(graph.serialize(format=format, encoding='utf-8'))

# Create ontology graph
ontology_graph = create_ontology()

# Serialize graph to a desired format (e.g., Turtle)
serialized_ontology = serialize_ontology(ontology_graph,
↪    format='turtle')

print(serialized_ontology)
```

This code defines a basic ontology for genetic data using several RDF concepts and constructs:

- `create_ontology` function constructs a simple RDF graph with genetic concepts including 'Gene', 'Protein', and 'Disease'.

- The function also establishes properties like `encodes` and `hasName` to demonstrate relationships within the ontology.

- Labels using `RDFS.label` for classes and properties are added for clarity.

- `serialize_ontology` function is used to output the ontology in common formats such as Turtle or XML, facilitating sharing and integration.

This ontology provides a foundational framework for organizing genetic data into a structured and shareable format, demonstrating the capability of semantic web technologies to support genetic data standardization.

Chapter 85

Natural Language Processing in Genomics

Below is a Python code snippet that encompasses the core computational elements of applying NLP techniques to extract information from biological literature, focusing on text parsing, named entity recognition, and relationship extraction using a library such as spaCy.

```
import spacy

# Load SpaCy's pre-trained NLP model
nlp = spacy.load('en_core_web_sm')

def extract_information(text):
    '''
    Extract entities and relationships from biological literature
    ↪ text.
    :param text: Input text from biological literature.
    :return: List of extracted entities and relationships.
    '''
    # Process the text with spaCy NLP model
    doc = nlp(text)

    entities = []
    relationships = []

    # Iterate over entities in the text
    for ent in doc.ents:
        entities.append((ent.text, ent.label_))

    # Iterate over the sentences to find token relationships
```

```
            for sent in doc.sents:
                for token in sent:
                    if token.dep_ == 'prep' and token.head.dep_ == 'ROOT':
                        # Example: Finding biological processes or
                        ↪   interactions
                        subject = [w for w in token.head.lefts if w.dep_ ==
                        ↪   'nsubj']
                        if subject:
                            relationships.append((subject[0].text,
                            ↪   token.text, token.head.text))

    return entities, relationships

# Example usage
if __name__ == "__main__":
    sample_text = (
        "The study demonstrates that the inhibition of the enzyme
        ↪   Phosphatase "
        "leads to increased growth in E. coli. Additionally,
        ↪   CRISPR-Cas9 was used "
        "to knockout the gene, leading to altered metabolic rates."
    )

    entities, relationships = extract_information(sample_text)

    print("Extracted Entities:")
    for entity in entities:
        print(f" - {entity[0]} ({entity[1]})")

    print("\nExtracted Relationships:")
    for relationship in relationships:
        print(f" - Subject: {relationship[0]}, Preposition:
        ↪   {relationship[1]}, Root: {relationship[2]}")
```

This code defines fundamental functions necessary for implementing natural language processing tasks on biological texts:

- `extract_information` function processes the input text to identify and extract named entities and illustrates basic relationship extraction between subjects and verbs in sentences.

- The use of spaCy's pre-trained NLP models simplifies linguistic feature extraction in complex biological literature.

- Sample text demonstrates how the code might be applied to real-world biological literature, highlighting potential relationships and significant biological entities such as enzymes or genes.

The final block of code provides examples of extracting entities and relationships, showcasing the application of NLP to decipher biological literature for insights into biological processes and interactions.

Chapter 86

Automated Hypothesis Generation in Biology

Below is a Python code snippet that encompasses the core computational elements for generating and testing biological hypotheses. It includes simulating hypotheses using random sampling, evaluating them through statistical tests, and updating the hypotheses based on results.

```python
import numpy as np
from scipy.stats import ttest_ind
import random

def generate_hypotheses(data, num_hypotheses=10):
    '''
    Generate a set of hypotheses based on random sampling of dataset
     indices.
    :param data: Data from which to generate hypotheses.
    :param num_hypotheses: Number of hypotheses to generate.
    :return: List of hypotheses, each represented as a tuple of data
     indices.
    '''
    hypotheses = []
    num_samples = len(data)
    for _ in range(num_hypotheses):
        sample_indices = random.sample(range(num_samples),
         num_samples // 2)
        hypotheses.append((sample_indices,
         list(set(range(num_samples)) - set(sample_indices))))
    return hypotheses

def evaluate_hypothesis(data, hypothesis):
```

```
'''
Evaluate a hypothesis by performing a t-test between two sample
↪  sets.
:param data: Data for hypothesis testing.
:param hypothesis: Tuple containing indices of two groups to
↪  compare.
:return: p-value from the t-test.
'''
    group1 = [data[i] for i in hypothesis[0]]
    group2 = [data[i] for i in hypothesis[1]]
    _, p_value = ttest_ind(group1, group2)
    return p_value

def update_hypotheses(hypotheses, data, significance_level=0.05):
    '''
    Update hypotheses based on statistical significance of results.
    :param hypotheses: List of hypotheses.
    :param data: Data against which to test hypotheses.
    :param significance_level: p-value threshold for hypothesis
    ↪  consideration.
    :return: Updated list of hypotheses considered statistically
    ↪  significant.
    '''
    accepted_hypotheses = []
    for hypothesis in hypotheses:
        p_value = evaluate_hypothesis(data, hypothesis)
        if p_value < significance_level:
            accepted_hypotheses.append(hypothesis)
    return accepted_hypotheses

# Example data and execution
np.random.seed(42)
data = np.random.randn(100)

# Generate initial hypotheses
initial_hypotheses = generate_hypotheses(data)

# Evaluate and update hypotheses
significant_hypotheses = update_hypotheses(initial_hypotheses, data)

print("Initial Hypotheses Count:", len(initial_hypotheses))
print("Significant Hypotheses Count:", len(significant_hypotheses))
```

This code defines several key functions necessary for the generation and testing of biological hypotheses computationally:

- **generate_hypotheses** creates randomized hypotheses by sampling indices of a given data array, which simulates different experimental scenarios.

- **evaluate_hypothesis** performs statistical hypothesis testing using a two-sample t-test to compare groups specified by a hypothesis.

- `update_hypotheses` checks each hypothesis against a significance criterion and retains those that show statistically significant results.

The final block of code provides examples of generating initial hypotheses from random data, evaluating these hypotheses using statistical methods, and retaining only those hypotheses that indicate significance. This structure applies common computational biology workflows to streamline hypothesis-driven research.

Chapter 87

Text Mining for Genetic Associations

Below is a Python code snippet that encompasses the core computational elements of extracting genetic associations from textual data, including text preprocessing, feature extraction, and using a machine learning model to identify associations.

```
import nltk
from sklearn.feature_extraction.text import TfidfVectorizer
from sklearn.linear_model import LogisticRegression
from sklearn.pipeline import make_pipeline
from nltk.corpus import stopwords
from nltk.tokenize import word_tokenize
import numpy as np

nltk.download('punkt')
nltk.download('stopwords')

def preprocess_text(text):
    '''
    Preprocess the input text by tokenizing, removing stop words,
    ↪ and stemming.
    :param text: Input string.
    :return: List of processed words.
    '''
    # Tokenize the text
    tokens = word_tokenize(text.lower())
    # Remove stop words
    filtered_tokens = [token for token in tokens if token not in
    ↪ stopwords.words('english')]
    return filtered_tokens
```

```python
def extract_features(corpus):
    '''
    Extracts TF-IDF features from the corpus.
    :param corpus: List of text documents.
    :return: Array of TF-IDF features.
    '''
    vectorizer = TfidfVectorizer(tokenizer=preprocess_text)
    return vectorizer.fit_transform(corpus)

def train_model(X, y):
    '''
    Train a logistic regression model to classify text data.
    :param X: Feature matrix from extracted features.
    :param y: Target labels.
    :return: Trained model.
    '''
    model = LogisticRegression(solver='liblinear')
    model.fit(X, y)
    return model

def predict_genetic_associations(model, text):
    '''
    Predict whether the text contains genetic associations.
    :param model: Trained model for prediction.
    :param text: New input text data for prediction.
    :return: Prediction score.
    '''
    # Assume the model is trained with a pipeline
    return model.predict_proba([text])[0][1]

# Example usage
corpus = ["Gene A interacts with Gene B", "Gene X is not associated
    with disease Y"]
labels = [1, 0]  # 1 indicates association, 0 indicates no
    association

# Extract features from the corpus
X = extract_features(corpus)
# Train the classifier model
model = train_model(X, labels)

# Predict for a new piece of text
new_text = "Gene B is associated with condition Z"
association_score = predict_genetic_associations(model, new_text)

print("Probability of genetic association:", association_score)
```

This code defines several key functions necessary for extracting genetic associations from text data:

- `preprocess_text` function processes the input text by removing stop words and tokenizing.

- `extract_features` retrieves the TF-IDF feature vectors from a given corpus.

- `train_model` fits a logistic regression model to classify the presence of genetic associations.

- `predict_genetic_associations` computes the probability that a given text implies genetic associations using the trained model.

The final block of code shows how to process a corpus, train a model, and predict associations using example text data.

Chapter 88

Visualization Techniques for Genomic Data

Below is a Python code snippet that encompasses the core computational elements for advanced visualization of complex genetic and genomic datasets. This script demonstrates data processing, dimensionality reduction, and interactive visualization using Python libraries.

```python
import pandas as pd
import numpy as np
import seaborn as sns
import matplotlib.pyplot as plt
from sklearn.decomposition import PCA
from sklearn.preprocessing import StandardScaler

def load_genomic_data(filename):
    '''
    Load genomic dataset from a CSV file.
    :param filename: Path to the CSV file containing genomic data.
    :return: DataFrame with normalized genomic data.
    '''
    data = pd.read_csv(filename)
    return data

def preprocess_data(data):
    '''
    Standardize the genomic data for PCA.
    :param data: DataFrame with genomic data.
```

```
    :return: Scaled data.
    '''
    features = data.columns
    scaler = StandardScaler()
    scaled_data = scaler.fit_transform(data)
    return pd.DataFrame(scaled_data, columns=features)

def perform_pca(data, n_components=2):
    '''
    Perform PCA transformation on genomic data.
    :param data: Scaled genomic data.
    :param n_components: Number of principal components.
    :return: DataFrame with PCA results.
    '''
    pca = PCA(n_components=n_components)
    principal_components = pca.fit_transform(data)
    pca_result = pd.DataFrame(data=principal_components,
    ↪   columns=['PC1', 'PC2'])
    return pca_result

def visualize_pca(pca_result, labels):
    '''
    Create a scatter plot for the first two principal components.
    :param pca_result: DataFrame with PCA results.
    :param labels: Cluster or sample labels for coloring.
    '''
    plt.figure(figsize=(10, 7))
    sns.scatterplot(x='PC1', y='PC2', hue=labels, data=pca_result,
    ↪   palette='viridis')
    plt.title('PCA of Genomic Data')
    plt.xlabel('Principal Component 1')
    plt.ylabel('Principal Component 2')
    plt.legend(loc='best', title='Sample Type')
    plt.show()

# Main execution
filename = 'genomic_data.csv'
raw_data = load_genomic_data(filename)
scaled_data = preprocess_data(raw_data)
pca_result = perform_pca(scaled_data)
sample_labels = raw_data['SampleType']
visualize_pca(pca_result, sample_labels)
```

This code defines several key functions necessary for analyzing and visualizing genetic data:

- `load_genomic_data` loads and reads the genomic data from a CSV file into a pandas DataFrame.

- `preprocess_data` standardizes the data, which is crucial for PCA when data is of varying scales.

- `perform_pca` performs principal component analysis to reduce data dimensionality and yield key insights.

- `visualize_pca` generates a scatter plot for the first two principal components using Seaborn for visualization.

The final block of code demonstrates loading a dataset, standardizing the data, applying PCA, and visualizing the results, assisting in meaningful interpretations of complex genomic data.

Chapter 89

Interactive Genome Browsers Development

Below is a Python code snippet that demonstrates interactive exploration of genomic data through the development of a simple genome browser using popular libraries like Matplotlib for visualization and Pandas for data manipulation. This example primarily highlights interaction with genomic datasets to visualize genome features.

```
import pandas as pd
import matplotlib.pyplot as plt
from matplotlib.widgets import Slider, Button

# Sample genomic feature data
data = {
    "chromosome": ["chr1"] * 5,
    "start": [2000, 4000, 6000, 8000, 10000],
    "end": [3000, 5000, 7000, 9000, 11000],
    "gene": ["gene1", "gene2", "gene3", "gene4", "gene5"]
}

# Convert data to DataFrame
df = pd.DataFrame(data)

def plot_genomic_features(start, window_size=5000):
    '''
    Plot genomic features in the specified range.
    :param start: Start position for visualization.
    :param window_size: Range of base pairs to display.
    '''
```

```python
    filtered = df[(df['start'] >= start) & (df['end'] <= start +
    ↪  window_size)]

    plt.figure(figsize=(10, 2))
    for idx, row in filtered.iterrows():
        plt.plot([row['start'], row['end']], [1, 1],
        ↪  label=row['gene'], lw=3)

    plt.xlim(start, start + window_size)
    plt.ylim(0.5, 1.5)
    plt.xlabel('Position')
    plt.title(f'Genomic Features from {start} to {start +
    ↪  window_size}')
    plt.legend()
    plt.show()

start_pos = 0
plot_genomic_features(start_pos)

# Interactive slider implementation
fig, ax = plt.subplots()
plt.subplots_adjust(bottom=0.25)
plot_genomic_features(start_pos)

ax_slider = plt.axes([0.25, 0.1, 0.65, 0.03])
slider = Slider(ax_slider, 'Start', df['start'].min(),
↪  df['end'].max() - 5000, valinit=start_pos)

def update(val):
    ax.clear()
    plot_genomic_features(int(slider.val), 5000)

slider.on_changed(update)

reset_ax = plt.axes([0.8, 0.025, 0.1, 0.04])
button = Button(reset_ax, 'Reset', hovercolor='0.975')

def reset(event):
    slider.reset()

button.on_clicked(reset)

plt.show()
```

This code defines several key components for setting up an interactive genomic data visualization tool:

- `plot_genomic_features` function visualizes genomic features within a specified base pair range.
- `Slider` from Matplotlib provides interactivity, allowing users to explore different genome segments dynamically.

- `Button` is used to reset the slider to its initial state for ease of exploration.

This setup forms the basis of a simple yet powerful framework for interactive exploration and visualization of genomic datasets, enabling users to engage with and interpret genomic features effectively.

Chapter 90

Algorithmic Music Generation from Genetic Data

Below is a Python code snippet that demonstrates the conversion of genetic sequences into musical compositions using algorithmic techniques. The code achieves this by mapping nucleotide sequences to musical notes and rhythms.

```python
import random

# Mapping of nucleotide bases to musical notes
note_map = {
    'A': 'C',
    'T': 'D',
    'G': 'E',
    'C': 'F'
}

# Function to convert a DNA sequence to a melody
def dna_to_melody(dna_sequence):
    '''
    Converts a DNA sequence into a musical melody.
    :param dna_sequence: A string representing the DNA sequence.
    :return: A list of musical notes corresponding to the DNA
        sequence.
    '''
    melody = []
    for base in dna_sequence:
        if base in note_map:
            melody.append(note_map[base])
```

```
        return melody

# Function to assign rhythmic patterns to a melody based on random
↪  choice
def apply_rhythm(melody):
    '''
    Applies rhythmic patterns to a sequence of musical notes.
    :param melody: A list of musical notes.
    :return: A list of tuples representing notes with rhythm.
    '''
    rhythms = ['quarter', 'eighth', 'half']
    melody_with_rhythm = [(note, random.choice(rhythms)) for note in
    ↪  melody]
    return melody_with_rhythm

# Example DNA sequence
dna_sequence = "ATGCGTAGCTAG"

# Convert DNA sequence to melody
melody = dna_to_melody(dna_sequence)

# Apply rhythm to the melody
melody_with_rhythm = apply_rhythm(melody)

# Output the musical composition
for note, rhythm in melody_with_rhythm:
    print(f"Note: {note}, Rhythm: {rhythm}")
```

This code defines two primary functions for the conversion process:

- **dna_to_melody** function transforms a DNA sequence into a sequence of musical notes according to predefined mappings.

- **apply_rhythm** assigns a random rhythmic value to each note, enhancing the melody with temporal structure.

The example illustrates the mapping of a sample DNA sequence and demonstrates the resulting musical composition through a series of notes and rhythms.

Chapter 91

Fractal Patterns in Genomic Sequences

Below is a Python code snippet that encompasses the core computational elements required to analyze and visualize fractal patterns within DNA sequences. This example demonstrates generating a graphical representation of such patterns using the Chaos Game Representation (CGR).

```python
import numpy as np
import matplotlib.pyplot as plt

def chaos_game_representation(sequence):
    '''
    Generate Chaos Game Representation for a given DNA sequence.
    :param sequence: DNA sequence string composed of A, T, C, G.
    :return: 2D numpy array representing the fractal pattern.
    '''
    # Define the vertices of the square for each nucleotide
    vertex_map = {
        'A': np.array([0.0, 0.0]),
        'T': np.array([1.0, 0.0]),
        'C': np.array([0.0, 1.0]),
        'G': np.array([1.0, 1.0]),
    }

    # Initialize the list of points in the plot
    x, y = 0.5, 0.5
    points = [(x, y)]

    # Iterate through the sequence and calculate positions
    for nucleotide in sequence.upper():
```

```python
            if nucleotide in vertex_map:
                vertex = vertex_map[nucleotide]
                x = (x + vertex[0]) / 2.0
                y = (y + vertex[1]) / 2.0
                points.append((x, y))

    return np.array(points)

def plot_cgr(points):
    '''
    Plot the Chaos Game Representation of the DNA sequence.
    :param points: 2D numpy array representing the fractal pattern.
    '''
    plt.figure(figsize=(8, 8))
    plt.scatter(points[:, 0], points[:, 1], s=0.1, c='blue')
    plt.title('Chaos Game Representation of DNA Sequence')
    plt.xlabel('X-axis')
    plt.ylabel('Y-axis')
    plt.show()

# Example DNA sequence
dna_sequence = "ATGCGTAGCTAGCTAGCTGACTGATCGTATGCGTAACTG"
points = chaos_game_representation(dna_sequence)
plot_cgr(points)
```

This code defines several key functions necessary for generating and visualizing fractal patterns in DNA sequences:

- `chaos_game_representation` function converts a DNA sequence into a set of points representing a fractal pattern using Chaos Game Representation.

- `plot_cgr` visualizes the generated fractal pattern from the calculated points.

The final block of code demonstrates plotting the Chaos Game Representation for a sample DNA sequence.

Chapter 92

Chaos Theory Applications in Genetics

Below is a Python code snippet that implements a basic simulation to model genetic regulatory networks using principles from chaos theory. This example demonstrates how chaotic behaviors can emerge in regulatory networks, using a simplified logistic map usually deployed in chaos theory studies.

```
import numpy as np
import matplotlib.pyplot as plt

def logistic_map(x, r):
    '''
    Logistic map function, a classic example of how chaotic behavior
      can arise
    from simple non-linear dynamical equations.
    :param x: Current state of the system.
    :param r: Control parameter for the map.
    :return: Next state of the system.
    '''
    return r * x * (1 - x)

def simulate_genetic_regulatory_network(r, x0, num_iterations):
    '''
    Simulate a genetic regulatory network using the logistic map.
    :param r: Parameter controlling the behavior of the system.
    :param x0: Initial state of the system (usually a small non-zero
      positive number).
```

```
    :param num_iterations: Number of iterations to simulate.
    :return: Array representing the state of the system at each
    ↪  iteration.
    '''
    states = np.zeros(num_iterations)
    x = x0

    for i in range(num_iterations):
        x = logistic_map(x, r)
        states[i] = x

    return states

# Parameters for simulation
r = 3.9  # Parameter value within the chaotic regime
x0 = 0.5  # Initial condition
num_iterations = 100

# Run simulation
states = simulate_genetic_regulatory_network(r, x0, num_iterations)

# Plot the results
plt.figure(figsize=(10, 6))
plt.plot(states, label='Genetic Regulatory Network State')
plt.title('Simulation of Genetic Regulatory Network using Chaos
↪  Theory')
plt.xlabel('Iteration')
plt.ylabel('State')
plt.legend()
plt.show()
```

This code defines the necessary components for simulating a genetic regulatory network influenced by chaotic behavior using the logistic map:

- `logistic_map` function computes the next iteration of the genetic regulatory system using a simple nonlinear dynamical system known for displaying chaotic properties.

- `simulate_genetic_regulatory_network` runs the simulation over several iterations, applying the logistic map to model the genetic regulatory network.

- The example uses initial conditions and a parameter (`r`) known to produce chaotic behavior, typical in explorations of chaos theory.

The plotting section at the end visualizes the change in state across iterations, showing how chaotic dynamics might manifest in a regulatory network model, reflected in wildly oscillating states.

Chapter 93

Multi-Objective Optimization in Genetic Engineering

Below is a Python code snippet that encompasses the core computational elements of multiple objective optimization in genetic design, such as defining objective functions, implementing an optimization algorithm, and balancing trade-offs between different objectives.

```python
import numpy as np
from scipy.optimize import minimize

def objective1(x):
    '''
    Objective function 1: Minimize cost associated with genetic
      modifications.
    :param x: Design variables.
    :return: Cost value.
    '''
    return 2 * x[0] + 3 * x[1] + x[2]**2

def objective2(x):
    '''
    Objective function 2: Maximize efficiency of the genetic
      circuit.
    :param x: Design variables.
    :return: Negative efficiency value (since we minimize).
    '''
    return -(x[0] * x[1] * x[2])
```

```python
def constraint1(x):
    '''
    Constraint 1: Ensures total resource usage is less than a
    ↪    threshold.
    :param x: Design variables.
    :return: Constraint value.
    '''
    return 10 - (x[0] + x[1] + x[2])

def constraint2(x):
    '''
    Constraint 2: Bounds genetic component rates to biological
    ↪    limits.
    :param x: Design variables.
    :return: Constraint value.
    '''
    return x[0] * x[1] - 1

# Initial guess for the design variables
x0 = [1.0, 1.0, 1.0]

# Define the constraints dictionary
cons = [{'type': 'ineq', 'fun': constraint1},
        {'type': 'ineq', 'fun': constraint2}]

# Combine multiple objectives into a single function using weights
def aggregate_objectives(x, w1, w2):
    '''
    Combine the objectives into a single objective function.
    :param x: Design variables.
    :param w1: Weight for the first objective.
    :param w2: Weight for the second objective.
    :return: Weighted sum of objectives.
    '''
    return w1 * objective1(x) + w2 * objective2(x)

# Weights for the objectives
w1, w2 = 0.5, 0.5

# Perform the optimization
result = minimize(aggregate_objectives, x0, args=(w1, w2),
    ↪    constraints=cons)

# Outputs for demonstration
print("Optimal Design Variables:", result.x)
print("Optimal Cost:", objective1(result.x))
print("Optimal Efficiency:", -objective2(result.x))
```

This code defines several key functions necessary for the implementation and evaluation of multiple objective optimization in genetic design:

- `objective1` defines an objective function that models the cost associated with genetic modifications, which should be minimized.

- `objective2` defines another objective function for maximizing the efficiency of a genetic circuit, implemented as its negative because optimization routines typically minimize objectives.

- `constraint1` and `constraint2` enforce constraints on the optimization problem, ensuring feasible solutions by limiting resources and respecting biological limits.

- `aggregate_objectives` combines multiple objectives into a single composite objective using specified weights for balancing trade-offs.

The final block of code demonstrates the use of these functions by performing optimization with initial guesses and constraint satisfaction, providing optimal design variables, cost, and efficiency values.

Chapter 94

Adaptive Algorithms in Evolutionary Biology

Below is a Python code snippet that implements adaptive methods to model evolution and natural selection computationally, including a genetic algorithm framework, fitness evaluation, selection, mutation, and crossover mechanisms.

```
import random
import numpy as np

def fitness_function(individual):
    '''
    Evaluate the fitness of an individual.
    :param individual: A list representing an individual's
    ↪ characteristics.
    :return: Fitness score.
    '''
    # Simple fitness function, e.g., sum of the individual's genes
    return sum(individual)

def create_individual(length):
    '''
    Generate a random individual.
    :param length: The number of genes.
    :return: A new individual.
    '''
    return [random.randint(0, 1) for _ in range(length)]

def mutate(individual, mutation_rate):
    '''
    Mutate an individual's genes based on a mutation rate.
    :param individual: An individual.
```

```
    :param mutation_rate: Probability of mutation per gene.
    :return: Mutated individual.
    '''
    return [gene if random.random() > mutation_rate else 1 - gene
    ↪   for gene in individual]

def crossover(parent1, parent2):
    '''
    Perform crossover between two parents to produce offspring.
    :param parent1: First parent individual.
    :param parent2: Second parent individual.
    :return: Two offspring.
    '''
    point = random.randint(1, len(parent1) - 1)
    return parent1[:point] + parent2[point:], parent2[:point] +
    ↪   parent1[point:]

def selection(population, fitnesses, num_to_select):
    '''
    Select individuals based on weighted fitness.
    :param population: Current population of individuals.
    :param fitnesses: Corresponding fitness scores.
    :param num_to_select: Number of individuals to select.
    :return: Selected individuals.
    '''
    total_fitness = sum(fitnesses)
    probabilities = [f / total_fitness for f in fitnesses]
    return random.choices(population, weights=probabilities,
    ↪   k=num_to_select)

def genetic_algorithm(population_size, gene_length, generations,
↪   mutation_rate):
    '''
    Run the genetic algorithm.
    :param population_size: Number of individuals in the population.
    :param gene_length: Length of the individual gene sequence.
    :param generations: Number of generations to iterate.
    :param mutation_rate: Mutation rate per gene.
    :return: Best individual and its fitness score.
    '''
    # Create initial population
    population = [create_individual(gene_length) for _ in
    ↪   range(population_size)]

    for generation in range(generations):
        # Evaluate fitness
        fitnesses = [fitness_function(individual) for individual in
        ↪   population]

        # Selection
        selected = selection(population, fitnesses, population_size
        ↪   // 2)
```

```
# Crossover
offspring = []
while len(offspring) < population_size:
    parent1, parent2 = random.sample(selected, 2)
    child1, child2 = crossover(parent1, parent2)
    offspring.extend([child1, child2])

# Mutation
population = [mutate(individual, mutation_rate) for
    individual in offspring]

# Evaluate final population
fitnesses = [fitness_function(individual) for individual in
    population]
best_index = np.argmax(fitnesses)

return population[best_index], fitnesses[best_index]

# Parameters
population_size = 100
gene_length = 10
generations = 50
mutation_rate = 0.01

# Run genetic algorithm
best_individual, best_fitness = genetic_algorithm(population_size,
    gene_length, generations, mutation_rate)
print("Best Individual:", best_individual)
print("Best Fitness:", best_fitness)
```

This code provides a comprehensive implementation of a genetic algorithm to model evolution computationally:

- `fitness_function` evaluates the fitness of an individual based on its genetic composition.

- `create_individual` generates a new random individual with a specified gene length.

- `mutate` alters the genes of an individual according to a specified mutation rate, introducing diversity.

- `crossover` combines pairs of individuals to produce new offspring with mixed genetic traits.

- `selection` uses a probabilistic approach to choose individuals for breeding based on fitness scores.

- `genetic_algorithm` encapsulates the iterative process of population evolution through selection, crossover, and mutation across generations.

The final part of the code executes the genetic algorithm, identifying the best individual and fitness score after all iterations.

Chapter 95

Information Theory in Genetics

Below is a Python code snippet that calculates information entropy and mutual information to analyze genetic data properties, exploring key elements of computational information theory applications in genetics.

```
import numpy as np
from collections import Counter
from sklearn.metrics import mutual_info_score

def calculate_entropy(sequence):
    '''
    Calculate the Shannon entropy of a genetic sequence.
    :param sequence: The genetic sequence (string or list) whose
    ↪    entropy is to be calculated.
    :return: Shannon entropy value.
    '''
    # Count frequency of each symbol in the sequence
    freq_dict = Counter(sequence)
    total_count = len(sequence)

    # Calculate entropy
    entropy = -sum((count/total_count) * np.log2(count/total_count)
                    for count in freq_dict.values())
    return entropy

def mutual_information(seq1, seq2):
    '''
    Calculate mutual information between two genetic sequences.
    :param seq1: First genetic sequence (string or list).
    :param seq2: Second genetic sequence (string or list).
```

```
    :return: Mutual information value.
    '''
    return mutual_info_score(seq1, seq2)

# Example genetic sequences
seq_a = "AGCTAGCTAGCTA"
seq_b = "TCTGCGTATGCCA"

# Calculate entropy for each sequence
entropy_a = calculate_entropy(seq_a)
entropy_b = calculate_entropy(seq_b)

# Calculate mutual information between the sequences
mi_ab = mutual_information(seq_a, seq_b)

print("Entropy of sequence A:", entropy_a)
print("Entropy of sequence B:", entropy_b)
print("Mutual Information between sequences A and B:", mi_ab)
```

This code defines key functions for analyzing genetic data using information theory:

- calculate_entropy computes the Shannon entropy of a genetic sequence, quantifying its uncertainty or information content.

- mutual_information employs the sklearn library to calculate the mutual information between two sequences, indicating the amount of shared information.

The example sequences illustrate the calculation of entropy for individual sequences and mutual information between two distinct genetic sequences.

Chapter 96

Game Theory Models in Genetic Interactions

Below is a Python code snippet illustrating the application of game-theoretic approaches to model competitive and cooperative interactions in genetics. This example uses fundamental concepts of game theory, such as payoff matrices and strategy formulation, to simulate interactions among genetic entities.

```
import numpy as np
import itertools

def build_payoff_matrix(interactions):
    '''
    Construct a payoff matrix from interaction data.
    :param interactions: Dictionary with interactions and associated
    ↪ payoffs.
    :return: Payoff matrix as a 2D numpy array.
    '''
    entities = list(interactions.keys())
    num_entities = len(entities)
    payoff_matrix = np.zeros((num_entities, num_entities))
    for i, entity in enumerate(entities):
        for j, opponent in enumerate(entities):
            if (entity, opponent) in interactions:
                payoff_matrix[i, j] = interactions[(entity,
                ↪ opponent)]
    return payoff_matrix, entities

def nash_equilibrium(payoff_matrix):
    '''
    Calculate the Nash equilibrium for a given payoff matrix.
```

```python
    :param payoff_matrix: 2D numpy array representing the payoff
    ↪   matrix.
    :return: Strategies for players at Nash equilibrium.
    '''
    num_strategies = payoff_matrix.shape[0]
    best_responses = []

    for player in range(num_strategies):
        # Comparing the row of the player to find the best response
        best_response = np.argmax(payoff_matrix[player])
        best_responses.append(best_response)

    return [tuple(np.eye(num_strategies)[i]) for i in
    ↪   best_responses]

def evolutionary_dynamics(initial_population, payoff_matrix,
↪   iterations=100):
    '''
    Simulate evolutionary dynamics over a number of iterations.
    :param initial_population: List representing initial population
    ↪   distribution.
    :param payoff_matrix: 2D numpy array representing the payoff
    ↪   matrix.
    :param iterations: Number of iterations for simulation.
    :return: Final population distribution.
    '''
    population = np.array(initial_population, dtype=float)
    num_strategies = payoff_matrix.shape[0]

    for _ in range(iterations):
        # Calculate fitness of each strategy
        fitness = payoff_matrix @ population
        # Adjust population distribution based on fitness
        mean_fitness = np.dot(population, fitness)
        population = population * fitness / mean_fitness
        # Normalize to maintain population proportions
        population /= sum(population)

    return population

# Example usage
interactions = {
    ('A', 'A'): 3,
    ('A', 'B'): 0,
    ('B', 'A'): 5,
    ('B', 'B'): 1
}

payoff_matrix, entities = build_payoff_matrix(interactions)
initial_population = [0.5, 0.5]  # Example: equal probability
↪   starting point
nash_eq_strategies = nash_equilibrium(payoff_matrix)
```

```
final_population = evolutionary_dynamics(initial_population,
↪   payoff_matrix)

print("Entities:", entities)
print("Payoff Matrix:\n", payoff_matrix)
print("Nash Equilibrium Strategies:", nash_eq_strategies)
print("Final Population Distribution:", final_population)
```

This code defines several critical functions to simulate and analyze genetic interactions using game theory:

- `build_payoff_matrix` constructs a payoff matrix from given interaction payoffs, where each entity's interactions are processed.

- `nash_equilibrium` calculates the Nash equilibrium, determining the optimal strategies for entities, indicating stable state scenarios.

- `evolutionary_dynamics` simulates the evolution of strategy frequencies over iterations, highlighting population changes due to strategic interactions.

With this framework, researchers can analyze how genetic interactions may stabilize or evolve over time, providing insights into cooperative and competitive dynamics in a genetic context.

Chapter 97

Artificial Life and Genetic Algorithms

Below is a Python code snippet that encompasses the core computational elements of simulating life-like behaviors using genetic algorithms in computational environments, including the creation of a population, evaluation of fitness, selection, crossover, and mutation processes.

```
import numpy as np
import random

def initialize_population(pop_size, genome_length):
    '''
    Initialize a population with random binary genomes.
    :param pop_size: Number of individuals in the population.
    :param genome_length: Length of each individual's genome.
    :return: List of binary genomes.
    '''
    return [np.random.randint(2, size=genome_length).tolist() for _
        in range(pop_size)]

def fitness_function(genome):
    '''
    Evaluate the fitness of a given genome.
    :param genome: Individual genome (list of 0s and 1s).
    :return: Fitness score (higher is better).
    '''
    return sum(genome)  # Fitness is the number of '1's in the
        genome.

def selection(population, fitness_scores, num_parents):
```

```
    '''
    Select a subset of the population based on fitness scores.
    :param population: List of current population genomes.
    :param fitness_scores: Corresponding fitness scores.
    :param num_parents: Number of parents to select.
    :return: List of selected parents.
    '''
    selected_indices = np.argsort(fitness_scores)[-num_parents:]
    return [population[i] for i in selected_indices]

def crossover(parent1, parent2):
    '''
    Perform single-point crossover between two parents to produce
    ↪  offspring.
    :param parent1: First parent genome.
    :param parent2: Second parent genome.
    :return: Two new offspring genomes.
    '''
    cp = random.randint(1, len(parent1) - 1)
    return parent1[:cp] + parent2[cp:], parent2[:cp] + parent1[cp:]

def mutation(genome, mutation_rate):
    '''
    Apply mutation on a genome based on the mutation rate.
    :param genome: Genome to mutate.
    :param mutation_rate: Probability of mutation for each gene.
    :return: Mutated genome.
    '''
    return [gene if random.random() > mutation_rate else 1-gene for
    ↪  gene in genome]

def genetic_algorithm(pop_size, genome_length, num_generations,
↪  mutation_rate):
    '''
    Run a genetic algorithm.
    :param pop_size: Number of individuals in the population.
    :param genome_length: Length of each genome.
    :param num_generations: Number of generations to simulate.
    :param mutation_rate: Probability of mutation for each gene.
    :return: Best genome and its fitness score.
    '''
    population = initialize_population(pop_size, genome_length)

    for generation in range(num_generations):
        fitness_scores = [fitness_function(genome) for genome in
        ↪  population]

        if generation % 10 == 0:  # Report every 10 generations
            print(f"Generation {generation}: Max Fitness =
            ↪  {max(fitness_scores)}")

        selected_parents = selection(population, fitness_scores,
        ↪  pop_size // 2)
```

```
        next_generation = []

        while len(next_generation) < pop_size:
            parent1, parent2 = random.sample(selected_parents, 2)
            offspring1, offspring2 = crossover(parent1, parent2)
            next_generation.extend([mutation(offspring1,
            ↪    mutation_rate),
                                    mutation(offspring2,
                                    ↪    mutation_rate)])

        population = next_generation[:pop_size]

    best_genome = max(population, key=fitness_function)
    best_fitness = fitness_function(best_genome)
    return best_genome, best_fitness

# Run the genetic algorithm
best_genome, best_fitness = genetic_algorithm(pop_size=100,
↪    genome_length=20, num_generations=100, mutation_rate=0.01)
print("Best Genome:", best_genome)
print("Best Fitness:", best_fitness)
```

This code defines several key functions required for simulating life-like behaviors using genetic algorithms:

- **initialize_population** generates a random initial population of binary genomes.

- **fitness_function** evaluates how well each genome meets the desired criteria.

- **selection** chooses the fittest individuals from the current population to become parents for the next generation.

- **crossover** combines pairs of parent genomes to produce offspring with traits from both parents.

- **mutation** introduces random changes to genomes to maintain genetic diversity within the population.

- **genetic_algorithm** brings all these components together to simulate evolution over multiple generations and determine the fittest genome.

The final block of code uses these functions to run a genetic algorithm with specified parameters, printing out the best genome and its fitness score at the end of the simulation.

Chapter 98

Synthetic Evolution and In Silico Selection

Below is a Python code snippet that demonstrates the computational aspects of simulating synthetic evolution and selecting desirable traits using evolutionary algorithms. This code includes setup for population initialization, evolutionary operations, fitness evaluation, and selection process.

```
import numpy as np
import random

def initialize_population(pop_size, gene_length):
    '''
    Initialize a random population.
    :param pop_size: Number of individuals in the population.
    :param gene_length: Length of each individual's genetic
    ↪ sequence.
    :return: A population of random genetic sequences.
    '''
    return np.random.randint(2, size=(pop_size, gene_length))

def fitness_function(genetic_sequence):
    '''
    Evaluate the fitness of a genetic sequence.
    :param genetic_sequence: An array representing the genetic
    ↪ sequence.
    :return: Fitness score (higher is better).
    '''
    # As an example, fitness can measure the number of '1's in the
    ↪ sequence
    return np.sum(genetic_sequence)
```

```python
def select_parents(population, fitnesses, num_parents):
    '''
    Selects the most fit individuals to be parents.
    :param population: Current population of genetic sequences.
    :param fitnesses: Fitness score array of the population.
    :param num_parents: Number of parents to select.
    :return: Array of selected parent genetic sequences.
    '''
    parents_indices = np.argsort(fitnesses)[-num_parents:]
    return population[parents_indices]

def crossover(parents, offspring_size):
    '''
    Perform crossover to produce offspring.
    :param parents: Array of parent genetic sequences.
    :param offspring_size: Number of offspring to produce.
    :return: Offspring population.
    '''
    offspring = np.empty((offspring_size, parents.shape[1]))
    crossover_point = np.uint8(offspring.shape[1]/2)

    for k in range(offspring_size):
        parent1_idx = k % parents.shape[0]
        parent2_idx = (k+1) % parents.shape[0]
        offspring[k, 0:crossover_point] = parents[parent1_idx,
        ↪   0:crossover_point]
        offspring[k, crossover_point:] = parents[parent2_idx,
        ↪   crossover_point:]
    return offspring

def mutation(offspring_crossover, mutation_rate=0.01):
    '''
    Apply mutation to the offspring.
    :param offspring_crossover: Offspring population after
    ↪   crossover.
    :param mutation_rate: Mutation rate.
    :return: Mutated offspring population.
    '''
    for idx in range(offspring_crossover.shape[0]):
        if random.random() < mutation_rate:
            mutation_point = 
            ↪   np.random.randint(offspring_crossover.shape[1])
            offspring_crossover[idx, mutation_point] = \
            1 - offspring_crossover[idx, mutation_point]
    return offspring_crossover

def evolve(pop_size, gene_length, num_generations,
↪   num_parents_mating, mutation_rate):
    '''
    Run the evolutionary algorithm to simulate synthetic evolution.
    :param pop_size: Initial population size.
    :param gene_length: Length of genetic sequence.
```

```
:param num_generations: Number of generations to simulate.
:param num_parents_mating: Number of parents selected for
↪   producing offspring.
:param mutation_rate: Mutation rate applied to offspring.
:return: The best genetic sequence evolved.
'''
population = initialize_population(pop_size, gene_length)
for generation in range(num_generations):
    fitnesses = np.array([fitness_function(ind) for ind in
    ↪   population])
    parents = select_parents(population, fitnesses,
    ↪   num_parents_mating)
    offspring_crossover = crossover(parents,
    ↪   offspring_size=pop_size-parents.shape[0])
    offspring_mutation = mutation(offspring_crossover,
    ↪   mutation_rate)
    population[0:parents.shape[0], :] = parents
    population[parents.shape[0]:, :] = offspring_mutation
return population[np.argmax([fitness_function(ind) for ind in
↪   population])]

# Example parameters for running the evolutionary algorithm
best_sequence = evolve(
    pop_size=100,
    gene_length=10,
    num_generations=50,
    num_parents_mating=20,
    mutation_rate=0.01
)

print("Best sequence evolved:", best_sequence)
```

This code snippet includes essential components for simulating evolutionary processes to select desirable traits:

- `initialize_population` function generates an initial population of random genetic sequences.

- `fitness_function` evaluates the fitness of each sequence, a crucial step for selection.

- `select_parents` selects the top-performing sequences to act as parents for the next generation.

- `crossover` function produces offspring by combining genetic information from parent sequences.

- `mutation` introduces random genetic variation in the offspring to explore new genetic regions.

- `evolve` orchestrates the evolutionary process over several generations to derive the best sequence.

In this demonstration, parameters such as population size, genetic sequence length, number of generations, and mutation rate are configured to suit the simulation's requirements. The output of the algorithm is a genetic sequence with the highest fitness score achieved after a set number of generations.

Chapter 99

Cellular Automata Models in Genomic Pattern Formation

Below is a Python code snippet demonstrating how cellular automata can be used to simulate and analyze patterns in genomic organization.

```
import numpy as np
import matplotlib.pyplot as plt

def initialize_grid(size, init_pattern='random'):
    '''
    Initialize the cellular automata grid with the given pattern.
    :param size: Size of the grid (n x n).
    :param init_pattern: Initial pattern for the grid; 'random' for
    ↪ random initialization.
    :return: Initialized grid as a numpy array.
    '''
    if init_pattern == 'random':
        return np.random.choice([0, 1], size=(size, size))
    else:
        grid = np.zeros((size, size))
        pattern_length = len(init_pattern)
        grid[size//2, size//2 - pattern_length//2:size//2 +
        ↪ pattern_length//2] = np.array([int(x) for x in
        ↪ init_pattern])
        return grid

def get_neighbourhood(grid, x, y):
    '''
```

```
    Get the neighborhood states around a specific cell (x, y).
    :param grid: Current grid state.
    :param x: X-coordinate of the cell.
    :param y: Y-coordinate of the cell.
    :return: List of states for the neighboring cells.
    '''
    return [grid[(x-1)%grid.shape[0], y], grid[x,
     ↪  (y-1)%grid.shape[1]],
            grid[(x+1)%grid.shape[0], y], grid[x,
     ↪  (y+1)%grid.shape[1]]]

def update_grid(grid, rule):
    '''
    Update the grid based on the defined rule for cellular automata.
    :param grid: Current grid state.
    :param rule: Transition rule as a function.
    :return: Updated grid state.
    '''
    size = grid.shape[0]
    new_grid = np.copy(grid)
    for x in range(size):
        for y in range(size):
            neighborhood = get_neighbourhood(grid, x, y)
            new_grid[x, y] = rule(grid[x, y], sum(neighborhood))
    return new_grid

def conway_rule(state, alive_neighbors):
    '''
    Conway's Game of Life rule for cellular automata.
    :param state: Current state of the cell (0 or 1).
    :param alive_neighbors: Number of alive neighboring cells.
    :return: New state of the cell (0 or 1).
    '''
    if state == 1 and (alive_neighbors < 2 or alive_neighbors > 3):
        return 0
    elif state == 0 and alive_neighbors == 3:
        return 1
    return state

def run_simulation(size, steps, init_pattern='random',
 ↪  rule=conway_rule):
    '''
    Run the simulation for a specified number of steps.
    :param size: Size of the grid (n x n).
    :param steps: Number of steps to simulate.
    :param init_pattern: Initial pattern for the grid.
    :param rule: Transition rule for the grid cells.
    :return: None
    '''
    grid = initialize_grid(size, init_pattern)
    plt.figure(figsize=(10, 10))
    for step in range(steps):
        plt.clf()
```

```
        plt.title(f'Step {step}')
        plt.imshow(grid, cmap='binary')
        plt.pause(0.1)
        grid = update_grid(grid, rule)

# Running the simulation
run_simulation(size=50, steps=100, init_pattern='random')
```

This code provides a simplified framework for simulating cellular automata based on genomic organization patterns:

- `initialize_grid` sets up the initial grid configuration with specified patterns, such as random initialization.

- `get_neighbourhood` retrieves the surrounding cell states necessary for determining the cell's next state.

- `update_grid` modifies the grid in each iteration according to a specified rule.

- `conway_rule` offers an example using Conway's Game of Life rules to demonstrate how cellular automata rules influence cell states.

- `run_simulation` visualizes the evolution of the grid over a given number of steps, providing insights into pattern formations.

Using this framework, you can explore various cellular automata rules and initial conditions to model diverse patterns and behaviors in genomic sequences.